The
Rescue Company

The Rescue Company

RAY DOWNEY

Fire Engineering Books & Videos

Copyright © 1992 Fire Engineering Books & Videos
Park 80 West, Plaza Two, 7th floor,
Saddle Brook, New Jersey 07662

LIBRARY OF CONGRESS CATALOGING-IN-PUBLICATION DATA

Downey, Ray.
 The rescue company / Ray Downey.
  p. cm.
 ISBN 0-912212-25-X
 1. Life-saving at fires. I. Title.
TH9402.D68 1992
 628.9'25—dc20   91-42652

Printed in the United States of America

Book designed by Bernard Schleifer

This book is dedicated to Firefighter 6th grade Kevin C. Kane
of Ladder Company 110 and Firefighter 1st grade Alfred E. Ronaldson
of Rescue Company 3 of the City of New York Fire Department,
who made the supreme sacrifice in the line of duty during the year 1991.

To Kevin and Al: "Truly the bravest of the bravest."

ACKNOWLEDGMENT

SPECIAL ACKNOWLEDGMENT MUST BE given to the many talented and highly skilled firefighters I worked with during my 30-year career. Sincere gratitude to the officers and members of Rescue Company 2, past and present, who taught me everything they knew about rescue and who have made this book possible. This elite group of highly motivated and dedicated firefighters is what rescue is about.

To Tom Brennan, a good friend, who convinced me to put in writing everything I knew about rescue and got me started in the business.

To my wife, Rosalie, and the gang: Joe, Marie, Chuck, Ray, and Kathy, a great big thank you for your support, patience, understanding, and inspiration during the past few years of compiling, writing, and rewriting the manuscript.

CONTENTS

Contents

Part Three: Operations and Planning

INTRODUCTION

As we approach the twenty-first century, we find the fire service continuing to struggle to maintain pace with modern technology as it applies to firefighting. Modern day firefighters have had to meet the ever-challenging changes in plastics, fabrics, foam rubber, building materials, and construction configurations, to name a few. It can be said that firefighting truly can be considered a science. "Rescue" has met the ever-changing challenges encountered and has been singled out as a field of specialty in the fire service.

Since the formation of the first rescue company in 1915, rescue firefighters have had to overcome the complexities presented by emergencies ranging from motor vehicle extrications to major earthquake search and rescue. Much has been written about rescue in trade magazines, but few books have been dedicated specifically to rescue in the fire service. This book takes the reader from "what rescue is about" to major operational planning for rescue operations. It describes the duties, responsibilities, and operational guidelines for rescue personnel from the firefighter up to the incident commander. The rescue firefighter with the expertise and capability to operate special tools can be as important as the incident commander who skillfully utilizes resources to mitigate challenging potential disasters successfully.

During the past few years, rescuers have had to operate at major earthquakes; hurricanes; tornadoes; aviation disasters; train, rail, and bus accidents; confined-space rescues; mine and well incidents; explosions, hazardous-materials incidents; water-related incidents; and a type of incident that has been occurring more frequently in recent years—building collapses.

The number of conferences, training exercises, seminars, and workshops being offered on international, national, state, and local levels to address the urban search-and-rescue problems with which rescue personnel have had to cope have been increasing dramatically. These exercises have been focusing on ways to gather heavy rescue and technological expertise to overcome the problems associated with urban search and rescue and the obvious need for improved instrumentation, technologies, methodologies, training techniques, and data-information systems. Modern technology generally has increased the complexities of rescue operations, and specialized state-of-the-art rescue equipment continues to meet these challenges with tools and equipment such as listening and detecting devices, fiberoptics and search cameras, and powerful cutting equipment that will help to maintain the specialized skills needed to resolve successfully any type of rescue situation.

The 30 years of experience represented by this book guides the reader chapter by chapter on the components of rescue operations such as planning, training, operations, tools and equipment, and personnel. Recollections of rescue incidents provided the base that made this book a reality. This book presents strategies that enhance chances for successfully resolving rescue operations. They include developing a rescue operational plan, adapting and changing the plan when necessary, and utilizing available resources. These procedures, however, are only a few of the key ingredients for successful rescue operations discussed in this book.

ABOUT THE
AUTHOR

RAY DOWNEY BEGAN his fire service career in 1962, following in the footsteps of two older brothers. After finishing probationary firefighters school, he was assigned to Ladder Company 35 in the Lincoln Center area of New York City's West Side. Looking for more action, he transferred to Ladder Company 4, "The Pride of Midtown," and worked in the busy Times Square area before moving on to his next assignment as a firefighter in Rescue Company 2 in Brooklyn.

The rescue work accomplished and the firefighting experience gained while in Rescue Company 2 were valuable assets in motivating him to pursue the rank of lieutenant. In 1972 he was promoted to lieutenant and assigned to the busy Harlem section of Manhattan. After a few months of covering in various firehouses in Harlem, he was assigned to Engine Company 58 in the "Fire Factory," so named because of its heavy workload. For the next five years, he worked in Engine Company 58 and transferred across the floor to Ladder Company 26 for the last two years of his assignment in the Fire Factory.

In July 1977 he was promoted to captain and reassigned to Brooklyn. While a covering captain, he was detailed to the Division of Training as a lead instructor at the Probationary FireFighters School. He then was selected by the fire commissioner to form and organize Squad Company 1, a fully equipped engine company that carried a full complement of ladder company tools. The assignment included responding to all working fires in selected areas of Brooklyn.

He returned to the "rescue" to become commanding officer of Rescue Company 2, his current assignment, in June 1980. He is responsible for the overall administration, training, and operations of this highly specialized unit. Rescue Company 2 responds to all working fires, motor vehicle and industrial accidents, major emergencies, water incidents, and any unusual incidents in Brooklyn.

He has served on many national committees in the fire service. A charter member of the International Association of Fire Chiefs Urban Rescue and Structural Collapse Committee, he also is a member of the National Association of Search and Rescue, the Federal Emergency Management Agency Rescue Equipment Working Group, and the Technical Review Panel for the Urban Search and Rescue Task Forces, which will assist the federal government in handling disasters occurring in the United States. The author of numerous articles on rescue procedures, he travels around the country to lecture on rescue-related topics and is a member of the editorial advisory board of and contributing editor to *Fire Engineering* magazine. A graduate of the Suffolk Community College fire science program, he has been a member of the International Association of Firefighters and the New York State Fire Chiefs Association for 20 years.

He and his wife, Rosalie, to whom he has been married for 31 years, live in Deer Park, Long Island. They have five children: Joseph, Marie, Chuck, Ray, and Kathy. The fire service tradition continues with Joe and Chuck, who are members of the City of New York Fire Department. Joe is assigned to Squad Company 1 and Chuck to Engine Company 235, both in Brooklyn.

The
Rescue Company

PART ONE

Getting Started— the Rescue Company and Its People

1

WHAT RESCUE IS ABOUT

COURTESY S. SPAK

RESCUE. This term has many meanings in the fire service. Webster's dictionary defines it as follows: "To free from confinement, danger, or evil." Firefighters often use the word rescue to describe heroic acts such as saving a life: "Eighteen-month girl rescued after being trapped for 58 hours in an abandoned well." "Earthquake victim rescued after five days in cellar of collapsed office building." "Worker rescued after falling in gravel pit." "Firefighters rescue family of four from raging, early-morning house fire." We hear and read about these rescues regularly.

Some departments call the units that provide emergency medical and ambulance service rescue units or squads. These rescue calls account for 75 percent of the responses for some departments. Do not confuse these units with heavy rescue companies.

A rescue company—designated also as a heavy rescue or rescue squad in various parts of the country—combines firefighting with operations that involve extrication, search and rescue, hazardous-materials response, scuba diving, and other specialized activities. This is the type of rescue company discussed in this book.

The increase in the number of disasters such as building collapses; earthquakes (Figure 1.1), hurricanes, and tornadoes (Figure 1.2); and plane (Figure 1.3), train, bus, and car (Figure 1.4) crashes has created the need for units equipped with an extensive variety of special tools and equipment and staffed by specially trained firefighters who can handle any types of incidents as well as fires.

3

Figure 1.1. Freeway collapse after California earthquake. COURTESY M. NAVARRO

Figure 1.2. Tree felled by tornado crushes automobile. COURTESY J. REGAN

Figure 1.3. Air cargo crash at J. F. Kennedy Airport, New York City. COURTESY S. SPAK

Figure 1.4. Motor vehicle accident. COURTESY S. WILLIAMS

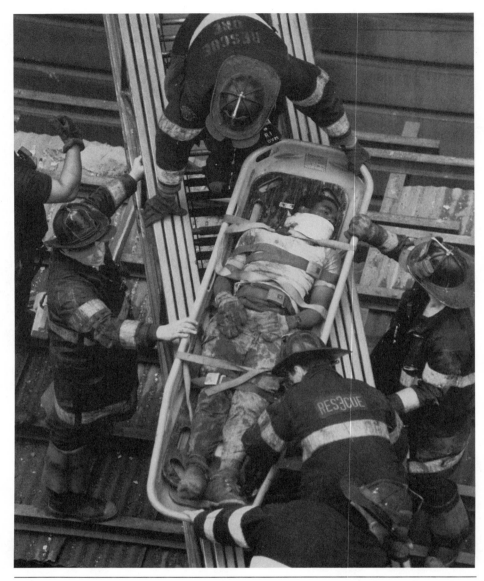

Figure 1.5. *Rescuers use aerial ladder to remove victim of a building collapse.* COURTESY
S. SPAK

Heavy rescue is performed by firefighters throughout the world.
Other emergency personnel may perform heavy rescue functions in some
parts of the world, but this situation is the exception rather than the
rule—and for good reason.

Firefighters on a daily basis deal with life-and-death circumstances
that require a diversified and thorough knowledge of building construction, designs and features, and elevators; hazardous materials; confined

Figure 1.6. Scuba rescuers remove victim to shore. COURTESY T. McCARTHY

space; and numerous other types of incidents that are part of the entry-level firefighting training.

Heavy rescue encompasses all the basic training of a firefighter *and* a specialization that is the result of additional distinct training that includes using and operating the equipment and tools required for heavy lifting, cutting (with everything from a hacksaw to a torch), below-grade and trench rescue, confined-space operations, vehicle extrication, search-and-rescue operations (Figure 1.5), and scuba operations (Figure 1.6).

Evolution of the Rescue Company

Three "auxiliary squad companies" were established in Chicago during 1913. They basically were manpower squads used to assist engine companies. It wasn't until 1929 that Chicago put into service three "rescue squads" whose specialized duties involved responding to accidents and inhalator runs.

On June 15, 1917, Boston's Rescue Company 1 was organized and placed in service. The company was equipped with smoke and gas hel-

Figure 1.7. *Members of the first rescue company don masks.* COURTESY J. CALDERONE COLLECTION

Figure 1.8. *Incidents such as this subway street collapse require specialized tools and equipment. The first rescue company was formed as a result of such unusual incidents firefighters face every day.* COURTESY J. CALDERONE COLLECTION

mets, a pullmotor, an elevator rescue outfit, oxygen- and acetylene-cutting devices, a 60-gallon chemical tank, hoses, axes, extinguishers, a life line, and other special tools and equipment. This company was organized to perform rescue work and to fight fires in inaccessible places, but it also was intended to perform other special work it might be assigned, such as responding to elevator rescues or chemical leaks.

The first "heavy duty rescue company" in the United States was placed into service in New York City on March 8, 1915, after the city had experienced a number of unusual and difficult fires in commercial buildings, piers, ships, and the underground subway system (Figure 1.7). This unit, consisting of a captain, a lieutenant, and eight firefighters, had among its equipment the following: smoke helmets, oxygen tanks, pullmotors, lungmotors, and a life line and gun. The members assigned to this unit were hand-picked, experienced tradesmen such as electricians, riggers, engineers, and mechanics trained in the use of the specialized tools and equipment for almost two months before being placed into service. The units performed ventilating and firefighting at fires and responded to a variety of rescue operations such as collapses, elevator rescues, chemical leaks, and other emergencies (Figure 1.8).

Rescue companies since their inception have carried tools and equipment, such as ropes, rigging, and lifting and cutting equipment, generally not carried by other fire department units. The technology that has evolved during the 75-plus years since the first U.S. heavy rescue unit was formed constantly has challenged the fire service to meet demands for more efficient, effective, and specialized tools and equipment and for more training in these tools and devices for heavy rescue company members. The training may be provided by manufacturers, the department, or an outside contractor.

Equipment

Although the amount and types of equipment a heavy rescue company carries are determined by the nature of the incidents the company most frequently encounters, several major categories of tools are likely to be found in any rescue company (Figure 1.9), including the following:

EXTRICATION EQUIPMENT

This category could range from hydraulic, air, or electrically operated spreading and pulling jaws to cutters to rams of various sizes and capabilities.

Figure 1.9. *Various tools and equipment carried by today's rescue companies.* COURTESY J. SKELSON

BURNING AND TORCH EQUIPMENT

These devices may be powered by oxygen or acetylene, or they could be specialized cutting torches powered by compressed air, batteries, or built-in igniting systems that utilize special rods for cutting. These exothermic cutting torches can be used to cut, burn, melt, or vaporize numerous types of materials.

JACKHAMMERS, ROTARY DRILLS, AND BREAKERS

These tools penetrate concrete, asphalt, and other materials. They also can be used horizontally for breaking brickwork, concrete, and similar materials.

SAWS

An assortment is needed: to cut wood, metal, and concrete.

ROPE, RIGGING, AND SHORING EQUIPMENT

(Figure 1.10).

Figure 1.10. Life-saving rope is part of a rescue company's tools and equipment inventory. Above, the rope is used to effect a dramatic rescue. COURTESY ASSOCIATED PRESS

GENERATORS

They must be able to supply power for lights and specialized tools.

METERS

These appliances are used to test explosive or toxic atmospheres as well as oxygen, carbon monoxide, and radiation levels.

HAZARDOUS-MATERIALS EQUIPMENT

Included are encapsulating suits, masks, in-line air systems, meters, and detectors.

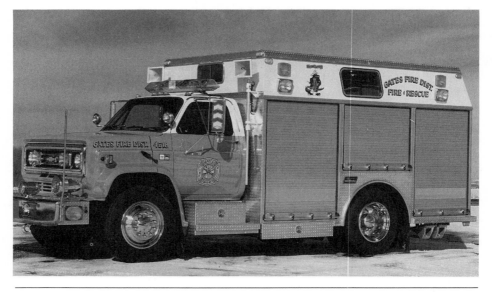

Figure 1.11. A smaller version of a heavy-rescue apparatus. COURTESY SAULSBURY APPARATUS COMPANY

SCUBA EQUIPMENT

This category includes wet suits, dry suits, tanks, regulators, gauges, and so forth.

The need to transport this huge array of tools accounts for the easily distinguishable design of rescue company apparatus. Its box-like shape is covered and filled with compartments for storing the tools in an orderly and systematic manner.

Heavy rescue equipment and apparatus vary according to the size and jurisdictional makeup of the department. State-of-the-art design now makes it possible for even small departments to carry enough equipment and tools to meet their needs in vehicles that provide sufficient compartment space (Figure 1.11). Most of these vehicles have a minimum carrying capacity of two rescue firefighters. The vehicles, however, can range from a large station wagon to a smaller version of a heavy rescue truck.

Personnel

The firefighters who work in a rescue company must be multitalented, highly experienced, and tremendously motivated. Their attitude must be, "Never give up."

Prior experience in construction or mechanical work often is a prerequisite for joining a rescue company. Previous training as a paramedic, emergency medical technician, or military medic also is helpful. Many rescue firefighters are military veterans who have had combat experience, and they tend to compare fighting the "red devil" with fighting the enemy. Just as their eventual assignment to a rescue company required out-of-the-ordinary skills, so, too, did their military backgrounds often include specialized functions such as those acquired as members of reconnaissance, airborne, special forces, or underwater-demolition teams.

Ideally, a rescue company should consist of six to seven firefighters, excluding the officer. Realistically, most companies ride with four to six members—three teams of two members each. Departments without the luxury of this type of staffing must adjust their normal response assignments and procedures to provide for rescue company responses. These options are discussed in Chapter 2.

Rescue-Related Responses

Although an incident commander may assign a rescue company to handle routine firefighting assignments such as ventilation, entry and search, and attack, the company's main purpose is to respond to unusual situations that other companies have neither the tools nor the expertise to handle (Figure 1.12). Following is a sampling of examples of some of these assignments:

AUTO ACCIDENTS

Spreaders, pullers, cutters, extending rams, jacks, blocks, and steering-wheel cutters—all may have to be used during the course of a single extrication. Doing all this while providing medical care for the victim is a team effort.

(During a building-collapse operation, for example, a former combat medic used his medical training and some psychological reassurance for two hours to stabilize a trapped construction worker as other rescue firefighters used their mechanical talents to free the victim.)

BUILDING COLLAPSE

The equipment required at the scene of a collapse can be a few hand tools or plus-power tools, specialized detecting devices, search cameras, search dogs, cranes, and other heavy-duty equipment.

Figure 1.12. A metal cutting saw is used to open gates. COURTESY W. FUCHS

Figure 1.13. This explosion caused a major fire and the collapse of two buildings. COURTESY H. EISNER

Collapses happen in many sizes and types of structures. I have responded to a variety of collapses. One incident involved a four-story, wood-frame building under renovation. Trench jacks, a reciprocating saw, and an entrenching tool were used to rescue a trapped worker (Figure 1.13). When a major fire and explosion caused the collapse of three occupied, two-story brick-and-joist buildings during another response, the equipment required included payloaders, forklifts, and trucks to remove the debris. Air bags, hand saws, small hand shovels, and a spreading and lifting tool—to free the victim—were used when a vacant, wood-frame structure collapsed; and the collapse of a six-story commercial building that killed one victim and trapped another for more than eight hours required a crane to remove sections of the building before rescuers could hand dig the victim out of the rubble.

ELEVATOR FIRES AND EMERGENCIES

Fires can occur in the elevator car itself, on top of the car, or in the elevator pit; or the fire could involve the cables or rails on which the elevator rides. Elevator emergencies could result from minor power losses or serious mechanical defects.

I once responded to an elevator emergency to find the building superintendent with his arm stuck between the elevator and the hallway shaft; he became trapped trying to reach in to trip the electrical contact to open the door. An air bag pushed the car back far enough to free his arm.

SUBWAYS AND TRAINS

Five cars of a 10-car train derailed and the first car was split open in a major train accident that occurred in New York City during August 1991. Five passengers were killed and 150 injured in the early-morning accident (Figure 1.14). Hundreds of firefighters, city and transit police officers, and medical personnel worked feverishly under extremely demanding conditions below the city's streets to free many of the injured trapped beneath the twisted wrecked cars. Among the hazards they faced while attempting the rescues were smoke, fumes, and concerns about the structural integrity of the subway tunnel, which had had one of its metal support columns forcibly struck by the derailing train.

A fire or emergency involving a railroad car in a tunnel presents the additional problems of finding a way to reach the scene, evacu-

Figure 1.14. *Train crashes are among the most difficult responses for rescue companies.*
COURTESY J. CALDERONE COLLECTION

ating crowds, stretching hose, avoiding electrified rails, maintaining communication, and working in smoke that reduces visibility to zero. Incidents with freight trains also may be complicated by the presence of hazardous materials. The number of rail incidents involving hazardous materials has been increasing dramatically, making it necessary to develop operating procedures for specifically handling these unusual incidents.

INDUSTRIAL ACCIDENTS

Disentangling a person from machinery is a complex job and not one for anyone who has a weak stomach. Many curious emergency workers crowded in on an operation to rescue a restaurant worker whose arm had been pulled into a meat grinder; as the rescue was nearing conclusion, many of them slowly backed away from the sight of the victim's injured arm.

MISCELLANEOUS TYPES OF RESCUES

Other operations might involve cave-ins, impalements, live electrical wires, hanging scaffolding, drownings, ice rescues, trench rescues, and rescues from manholes and other confined spaces. Some incidents are less dramatic and involve mundane activities and objects (Figure 1.15).

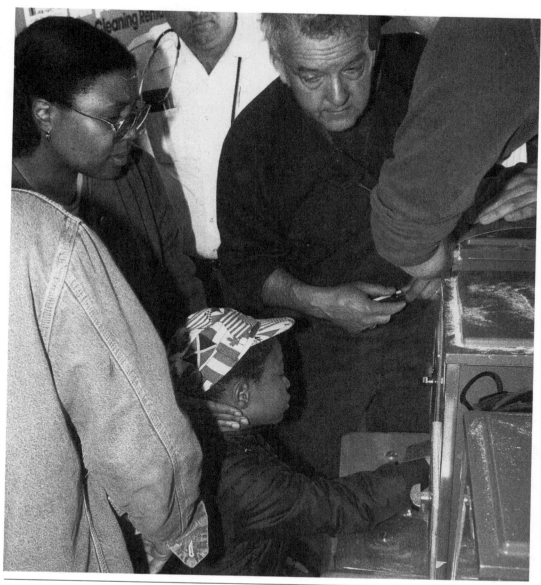

Figure 1.15. *Among rescue calls that come under the heading "different" is freeing a boy's hand from a gumball machine.* COURTESY S. SPAK

No two incidents are likely to be the same for a rescue company. Each event presents lessons to learn. Engaging in drills and incorporating lessons learned from past incidents are extremely important for ensuring that team members will be prepared to respond to a variety of rescue incidents. Critiquing the rescue operation and recording the information helps ensure that all the lessons learned will be included in the training sessions. As in all types of fires and emergencies, the pre-incident planning and training for rescue teams must be interwoven with good communication, coordination, control, and cooperation.

Selecting members for the rescue unit is a difficult task. A systematic and thorough approach can help to make it easier.

With safety as a primary concern, members must be used in a manner that maximizes the potential of their skills.

2

RECRUITING AND STAFFING

COURTESY OF RESCUE COMPANY 2.

Cʜᴀɴɢɪɴɢ ᴄᴏɴᴅɪᴛɪᴏɴꜱ ꜱɪɴᴄᴇ the first rescue company was established may have altered some aspects of recruiting, but the aura of "specialness" associated with being a member of a rescue company has not been dimmed by the passing of time.

The Recruitment Process

When one of the oldest rescue companies in New York City was being formed, the department had to recruit volunteers. Requirements included that applicants have some experience in a trade and, more importantly, pass a rigid examination. Because it was anticipated that they would be exposed to hazardous chemicals, these rescue firefighters also had to be in excellent health (those were pre-mask days).

More than 200 firefighters answered the call. The commanding officer of the first unit had the luxury of choosing the elite of the elite. Present-day circumstances probably would not allow the same luxury, but, as noted, the "specialness" attached to being a member of a rescue company has carried through to today. Membership in a rescue company is a special assignment; that in itself is enough to attract recruits who desire to be extraordinary.

How does a commanding officer choose from among a number of applicants? The answer lies in establishing a policy that fits the unit's needs. If the unit responds to hazardous-materials incidents in addition to its regular assignment, firefighters with some experience with hazardous materials would be needed, for example; and if the rescue company were responsible for water-rescue operations, its members certainly would be required to be qualified rescue divers.

The Selection Process

Following are some suggested guidelines for selecting rescue unit members:

THE INTERVIEW

The commanding officer, upon receiving a request for assignment to the rescue unit, should set up an interview with the firefighter. The face-to-face meeting should be private and conducted in a relaxed atmosphere (Figure 2.1). The interview is the mechanism that helps the commanding officer to determine whether the firefighter's interest is sincere or merely a "front-piece collector," and gives the firefighter the chance to ask questions. Either party's misconceptions should be corrected during the meeting. It's the ideal time for the commanding officer to explain the unit's functions, the company's policies, and the member's duties and responsibilities.

Figure 2.1. The face-to-face interview should be private and provide a relaxed atmosphere for the applicant. PHOTO BY AUTHOR

INTERVIEW FORM

RESCUE COMPANY_____ DATE_____

NAME_____ TEL. #_____

ADDRESS_____

COMPANY_____

DATE APPOINTED_____

DATE ASSIGNED TO PRESENT UNIT_____

LIST ALL PREVIOUS UNITS AND DATES SERVED:

     UNIT   DATES

LIST ALL SPECIAL QUALIFICATIONS: (TRADES-CONSTRUCTION, SPECIAL COURSES:
                                     EMT, AMT, PARAMEDIC, ETC.)

ON-THE-JOB   OFF-THE-JOB

ANY SPECIAL LICENSES, TRADES, OTHER:

ARE YOU ON A PROMOTIONAL LIST?

LIST_____LIST NUMBER_____

COMMENTS:

                                        _____
                                        INTERVIEWING OFFICER

Figure 2.2. The application form can be designed to provide the information that will meet any unit's special needs.

THE COMPANY APPLICATION FORM

The firefighter should fill out a special company form (Figure 2.2) that requests the following information:

- Date of appointment: To establish the firefighter's seniority standing in relation to the overall department
- Date assigned to the present unit: To determine the length and types of the applicant's experience while in that unit
- Previous companies and dates of assignment: To provide a background and experience checklist
- Prior experience on or off the job that would be beneficial to the rescue unit: To make the commanding officer aware of any special training or skill the applicant could bring to the unit (Figure 2.3)

Areas of expertise might include having had truck-company experience with power tools, extrication equipment, saws, and the like; hazmat training, if required; special department courses taken; scuba training, if required; and emergency medical technician (EMT), advanced medical technician (AMT), or paramedic training. Off-the-job expertise would include experience in construction-related trades such as a builder, carpenter, electrician, sandhog, rigger, torch operator, or welder.

Figure 2.3. Firefighters who can bring special skills to the unit should be considered.
COURTESY J. NORMAN

The form can be designed to provide information pertaining to any of the unit's special needs. Everyone would like to list the qualifications of a Rambo or James Bond, the multitalented firefighter with nine lives. The special application form, in conjunction with the interview, presents a more objective picture of the applicant.

Another factor to be considered is the firefighter's position on a promotion list. Depending on the number and variety of tools and equipment the applicant's unit has on its apparatus, it may take months or years to train a new rescue unit member to become familiar with and capable of operating all of the specialized equipment. Training a firefighter who may stay with the unit for only a short time adds to the rescue company's workload and deprives another applicant of a chance to be assigned to the unit. Applicants who have special skills to offer and who will be with the unit for an extended period of time should be given priority. If the field of candidates is limited, then adjustments consistent with the unit's needs will have to be made.

THE OFFICER'S REPORT

After the interview, the commanding officer should note on the interview form all pertinent comments and observations. The form then should be filed in a "Waiting List Book," from which the next *detail* would be selected. Detailing, as opposed to immediately transferring into a rescue unit, provides a period of time, usually one month to six months, for firefighters to see whether they really want to work in the rescue unit and for the commanding officer to see whether the firefighters can perform to expectations. Rescue work isn't everybody's cup of tea. If a department doesn't have provisions for detailing, then the firefighters should be given the option to return to their original units.

I recall hearing a story during my early firefighting days about a fellow firefighter with whom I was working. He was respected by all his peers for his dedication, motivation, and firefighting skills. When given the opportunity to apply for assignment to a rescue unit, he seized it. Some of his duties while detailed to the unit included responding to two major disasters—a major air crash and a disastrous fire aboard an aircraft carrier that was under renovation (Figure 2.4). He witnessed terrible devastation and loss of life. It's not very often that two major disasters occur so close together and that the same people have to perform the same jobs. At the end of his detail, he chose to return to his original unit. His decision was well-respected; it was easy to see that these two incidents had had a great effect on him.

Figure 2.4. *The above major disasters occurred within days of each other and tested the commitment of a young detailed rescue firefighter.* COURTESY J. CALDERONE

At times, "outstanding" firefighters may be recommended to commanding officers by other officers within the department. While this practice can be very helpful, the firefighter must want the assignment.

Utilizing Personnel

Today's fire service leaders, however, are faced with numerous problems, and staffing is a primary concern. Many of these leaders must deal with the "think tanks" of city administrations. The fire department usually is the first agency to be cut back during times of fiscal woes. A loss of one human life due to this type of fiscal insanity never can be fully understood or morally justified.

The Tenement Collapse

The following account of a building-collapse incident illustrates how the number of available workers can affect the outcome of operations. What started as a quiet summer morning tour for my unit hours later ended up testing its members' ability to handle a serious building collapse. A five-story tenement built in the late 1800s was being renovated to accommodate the influx of young professionals seeking reasonably priced living space. It marked the resurgence of an area that only a few years before unofficially had been declared dead.

The building had been completely gutted, and the new construction was to include poured concrete over a Q-decking. Workers were nearly finished pouring concrete on an upper floor when that floor suddenly collapsed. The concrete formwork was not properly installed to hold the concrete's excessive weight. The resulting collapse of lower floors sent workers scurrying in all directions. Fire department members arriving at the scene were met by a number of injured and disoriented construction workers who had problems remembering where they were; how many workers had been at the job site; what floor they had been on prior to the collapse; who, if anyone, was unaccounted for; and what actually happened.

Fortunately, the first-arriving chief calmly gathered and pieced together all the information available and was able to institute a plan of action. He had ascertained that while concrete was being poured on the third floor in an irregularly shaped area of the building, the floor suddenly began to vibrate and then quickly collapsed. A number of workers on that floor actually rode the collapsing floor down; although they suffered some injuries, they had been thrown clear of the rubble.

A worker on the outer edge and lead end of the collapsing floor was not so lucky. Remaining portions of the second and third floors had fallen in a lean-to position, trapping him against a bearing wall. The chief, realizing the seriousness of the situation, requested that a second rescue company respond (Figure 2.5).

The first rescue officer arriving at the scene conferred with the incident commander. After assessing the collapse conditions, he set an operational plan in action. His years of rescue experience told him that he was in for an extended operation. Would he and the five rescue firefighters, the department's staffing-level requirement, be enough?

The victim not only was trapped against a wall with portions of two floors pinning him, but he also had his leg trapped in an area that separated the lower portions of the collapse and cellar ceilings. Rescue workers would be working for long periods in areas that had the potential for further collapse. Relief and rotation would be priorities during these operations, and members would be required to work in front of, alongside, below, and above the trapped victim.

FIVE TWO-MEMBER TEAMS

With the arrival of the second rescue company, the rescue officer in charge was able to set up five two-member teams (Figure 2.6). Two of them worked with air bags and hydraulic spreading devices alongside the victim while another team shored the floor from below. The fourth team tried to free the victim's severely injured leg, and the fifth team secured the victim to a life line so that he wouldn't slip any farther down into the collapse rubble as rescuers worked to free him.

A delicate operation such as this one required precise coordination and teamwork from all involved. The expertise and personnel of a third rescue company had been summoned and placed in a backup position in case a secondary collapse should occur or the other rescue companies need help. How many departments today with their present staffing levels and number of available special units, such as rescue companies, would have been able to handle this type of operation?

Staffing Recommendations

What are the staffing recommendations for a rescue unit? The Municipal Fire Administration's recommendations for "Required Strength of Fire Companies" does not include staffing for rescue com-

Figure 2.5. A second company was needed to handle this collapse. COURTESY S. SPAK

Figure 2.6. The operations officer must consider setting up teams in a manner that will use resources most effectively. COURTESY H. EISNER

panies. The American Insurance Association's Bulletin No. 319 recommends six firefighters for all companies. It states: "Progressive fire chiefs are of the opinion that companies should never be allowed to respond with less (sic) than five members." Most nationally recognized fire organizations are of the same opinion. Some departments, depending on the nature of the fire district or its responses, may vary their staffing levels in certain companies.

SIX-MEMBER RESPONSE TEAM

Ideally, six firefighters and an officer should be the required staffing for all rescue companies (Figure 2.7). Unfortunately, economic hardships (the cry of most U.S. cities) or union contractual obligations have dropped the level down to five, and in some cases even four, firefighters per unit.

In the incident described above, 10 firefighters and two officers worked for more than two hours to free the victim. Fortunately, the department was able to provide the necessary personnel to ensure a successful rescue operation.

In February 1988, another large city was involved in a collapse similar to the one described above, but this time the trapped victims were firefighters. The problems were compounded when the trapped members turned out to be from the same special unit responsible for rescuing trapped victims—the only such unit in the department. Although the story had a happy ending, the department is taking steps to train another unit in this type of specialized work.

Regardless of the type of team formation, the firefighters must work together. They are taught from their first day in the department that teamwork is essential for successful firefighting. Whether it be forcible entry, raising ladders, or stretching hose, teamwork is the key for success (Figure 2.8). At most rescue operations, rescue firefighters provide the expertise to operate the specialized equipment. The teamwork required of rescue firefighters entails an extensive coordination of skills.

The "eyes and ears" of a two-member rescue team often are the factors that make an operation safe and successful. The guide, for example, can see, anticipate, and direct (often by hand signals), while the tool operator cannot do any of those things because hearing and sight often are obstructed by the high noise level and the size and weight of a tool or the nature of the working space. The rescue officer has the leverage to adapt the operational plan to the personnel available; the conditions being faced; and, most importantly, the expertise of the rescue team.

Figure 2.7. LEFT: *Ideally, six firefighters and an officer should be the required staffing for all rescue companies.* COURTESY S. SPAK

Figure 2.8. RIGHT: *Firefighters are taught from their first day in the department that teamwork is the ingredient essential for successful operations.* COURTESY W. FUCHS

In the collapse described at the beginning of this chapter, five two-member teams were required to free a trapped worker. One member of each team was designated as the "eyes and ears." For example, the tool operator of the air bags gently inflated the bags as directed by the guide, who could see the result of each movement of the bag. The operator of the hydraulic spreading device was directed in the same manner. This procedure enabled the rescue officer to supervise the overall operation by moving from team to team and noting the degree of progress.

An important advantage of having six members in a rescue company is that three highly trained and experienced two-member teams can provide an enormous amount of expertise in any given rescue operation. Their previous training and experience give them the knowledge and confidence to operate independently with the required team coordination.

Realistically, many rescue operations are undertaken with a fewer-than-ideal number of personnel, and working with five (instead of six) rescue firefighters and an officer can be successful in many cases. As is the case with fire operations, every rescue operation is different.

FIVE-MEMBER RESPONSE TEAM

In a five-member operation, the officer can set up two, two-member teams and work with the fifth person as the third team or assign the fifth member to a less demanding position. Using the fifth member as relief for the other two teams is another possibility.

Supervisors should be able to coordinate the rescue operation and, if necessary, to engage physically in it (Figure 2.9). Clearly, however, this situation should not be the "golden rule." Rescue operations are unique, and the officers directing them should have the expertise and capabilities to provide the effective coordination and communication needed to ensure success.

A five-person operation could be successful when rescue operations take place in a limited-access area. Only one two-person team can be used to approach the victim. The other teams may be used to survey other possible approaches, to work in other areas (for example, shoring

Figure 2.9. Supervisors should be able to coordinate the rescue operation and, if necessary, engage in it physically. COURTESY H. EISNER

from below or providing stability from above), or to act as the backup to the first team. The ways in which rescue firefighters can be used are limited only by the imagination of the rescue officer.

FOUR-MEMBER RESPONSE TEAM

A four-member team also can be effectively employed by the rescue officer. After sizing up and evaluating the conditions, the officer can implement an operational plan that uses two teams of two firefighters each or one team of three (with the fourth firefighter used for relief or special assignments). This arrangement allows the officer to supervise the operation and still be able to relay information and progress to the incident commander.

THREE-MEMBER RESPONSE TEAM

If the response is restricted to a three-member rescue unit, the officer can set up, as conditions warrant, the type of operational plan that best accommodates a three-person team. The officer, for example, could use one three-person team, and he/she could supervise and act as the fourth person, or the officer could use one two-member team and use the third firefighter for relief or special assignment.

As noted, the rescue officer has the leverage to adapt the operational plan to the personnel available; the conditions being faced; and, most importantly, the expertise of the rescue team.

Detailing affects the plan of operations and deployment of firefighter expertise. The rescue officer must be conscious of the detailing members in the company who are not trained in rescue and must adapt the plan accordingly. As an example, how would a rescue officer use a fourth firefighter detail with two years of experience in a fairly inactive engine company? More often than not, the person would be used in the nonrescue phase of the operation. Using the firefighter to transport and provide the tools and equipment as requested by the rescue team would be another effective alternative.

One of the most important resources a fire department has is the specially trained, multitalented, and highly motivated firefighters of its rescue companies. If a department must cut manpower, adequately staffing its rescue companies should be its utmost priority. Diluting a fine mixture leaves a bad taste in everyone's mouth.

Like rescue company personnel, rescue company apparatus must be carefully selected and especially suited for the rescue unit.

3

RESCUE APPARATUS

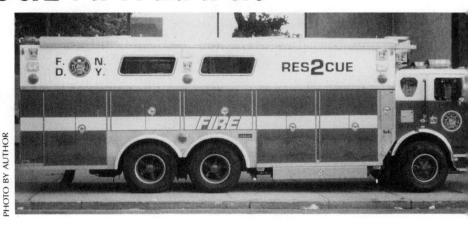

THE MANUFACTURERS OF rescue apparatus consistently have met the challenges the fire service has had to face as it approaches the twenty-first century. When we think of a rescue company, a variety of hydraulic-, air-, electric-, and gasoline-powered tools come to mind. We often overlook the apparatus, the piece of equipment that's the key to any rescue operation. It not only carries the specialized tools and equipment; but, more importantly, it's the means of transporting the personnel with the know-how. Purchasing, selecting, equipping, and designing the apparatus are the first steps toward successfully completing a rescue operation.

If you've ever attended an apparatus exposition or show, you've probably noticed that there is a variety of apparatus and that they differ in size and design. They may be custom-made or they can be adapted to the individual needs of a department. The Chicago Fire Department, for example, had its new rescue apparatus equipped with 55-foot snorkels. Many fire departments now order rescue apparatus with the specialized equipment needed for a hazardous-materials unit.

Mistakes Can Be Costly

What does a unit need? Who should be involved in the purchase, selection, and design of the apparatus for a department? I'm sure most of us in the fire service have heard about the department that spent many

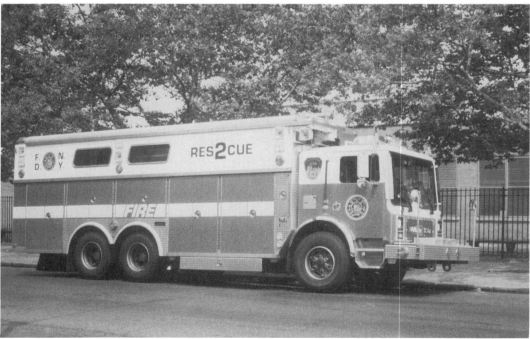

Figure 3.1. (a) 1931 Mack Bulldog, City of New York Fire Department. COURTESY S. SPAK. (b) 1990 Mack MR with Saulsbury body, City of New York Fire Department. PHOTO BY AUTHOR. (c) 1949 Mack, Chicago Fire Department. COURTESY J. REGAN. (d) 1987 Spartan-Emergency One, Chicago Fire Department. COURTESY J. REGAN

months planning, designing, and selecting a special piece of apparatus only to find when the delivery date arrived that the rig did not fit in its new home (Figure 3.1). The department did not know that the street on which the unit was stationed had been badly damaged by floods. Between the date the apparatus was purchased and the date it was delivered (which can be as long as two years), the street had been torn up and repaved. The apron to the firehouse had to be resloped to conform to the newly paved street. The apron's height didn't affect the apparatus already stationed within the firehouse, but when the driver tried to back in the shiny new apparatus, it wouldn't fit. Granted, this was an unusual situation, but the size of the station and any changes to it and its vicinity that might affect the department's use of new apparatus must be considered.

Other factors also must be considered if costly mistakes are to be avoided. Many of the older firehouses, for example, may not be able to withstand the additional weight or size of the newer rescue apparatus. Additional compartment space, built-in generators, cascade systems, and tandem wheels add extra weight that the apparatus floor will not support.

Another department's apparatus was designed with a very modern, customized bench seat that had a large storage space beneath the seating area. What was thought to be a neatly designed and highly efficient idea turned out to be a disaster. No thought had been given to the surrounding area, particularly aisle space. Department members quickly learned to walk in and out of the apparatus at a slight angle. Passing each other within the cab was impossible. The enticement of the large storage area overshadowed consideration of the day-to-day necessity of having a functional, fairly comfortable area.

Solicit Member Input

These examples bring home an important point: Enlisting the help of department members when selecting and designing apparatus makes good sense and could save major headaches. What better way to come up with ideas than to use the expertise of the members who drive and ride these rigs?

The first time I was to provide input relative to the design of a new apparatus, I was totally unaware of what to expect. I immediately enlisted the advice and ideas of the members of the unit. Surprisingly, the primary concern was the chassis. (If you have ever ridden in the back of a rescue apparatus and have had to take the bouncing around on every

Figure 3.2. *Apparatus designed for a department's needs must be full-sized and heavy-duty, such as the one shown here.* COURTESY SAULSBURY APPARATUS COMPANY

run, you would understand their concerns.) The apparatus assigned to the unit had a history of breaking rear springs; the weight of the apparatus and equipment was just too much for the chassis (Figure 3.2). The unit's veteran chauffeur had had experience purchasing heavy trucking for a commercial firm and suggested a heavy-duty rear end for the proposed apparatus. His suggestion was well received, and all members prided themselves on the fact that their involvement helped to eliminate "downtime" and the expensive, frequent spring changes.

When the new apparatus was delivered, however, we found that the manufacturer had installed a smaller rear end. When the powers that be were questioned, they responded that federal guidelines were used to spec out the new apparatus. Unfortunately, the creators of these guidelines had never considered the condition of the streets or the length, type, and frequency of this unit's responses.

Whether it's a formal apparatus selection committee or an informal group composed of members solicited for their ideas and input, its members should draw up a list of clearly defined needs the apparatus must fulfill for that unit. Career and volunteer departments can use the lessons they learned from previous purchases or the experiences of other departments. The members drawing up specifications for the new vehicle must be familiar with the unit's daily operations; the building in which the new apparatus will be housed; and, most importantly, the unit's needs.

What would the members consider a major need? Would it be the size and style of the gold leafing adorning the sides of the apparatus or the type and size of the unit used to heat the crew areas? Just ask any fire-fighter who has worked all night in subzero-degree temperatures which is more important. What may appear to be an item of small concern for the designer may be just the opposite for the department members.

BENCH OR SEATING AREAS

Benches and seating areas, for example, often aren't given the attention they deserve. A seating area must supply some comfort and also

---

Figure 3.3. *SCBA bracket designed to hold mask and allow officer to mask-up while responding.* PHOTO BY AUTHOR

incorporate the provisions that allow for immediate donning of self-contained breathing apparatus (SCBA). An apparatus delivered to one department had a comfortable backrest for the officers; but the lack of room in the cab forced provisions for the SCBA to be left out. The unit's officers couldn't afford to take the time to go to a compartment on the side of the apparatus while the unit's members were masked up and waiting to go. The backrest, therefore, was removed and replaced with two small pads, and an SCBA bracket was installed, making it possible for the officers to be masked and equipped in proper fashion (Figure 3.3). This unit's policy dictated that all members respond in mask equipment, another need that the designer hadn't considered.

Many departments are using the crew-cab design, in either the walk-through or separate-cab model, for rescue unit apparatus. Mask brackets can be incorporated into both models. Some planning can help create the system most appropriate for a department.

Special Compartments

The size and shape of the cab influence the overall length and compartments' spacing and sizes. The more compartments, the more equipment that can be carried. Many manufacturers of rescue apparatus build their units with specially designed compartments. Among these compartments are the following:

GENERATOR COMPARTMENT

This popular feature has built-in generators that supply power for portable lighting or electric tools and also ample power to provide electricity during blackouts or power shortages (Figure 3.4). The compartments generally carry a varied assortment of lights, electrical connections, and portable and reel-mounted extension lines. A large area can be illuminated by the lighting attached to and powered by the apparatus.

LIGHT TOWER COMPARTMENT

Many new apparatus have built-in light towers that provide air-operated extension systems capable of extending up to 40 feet high (Figure 3.5); the tower's rotating lights greatly enhance the illumination at rescue operations.

Figure 3.4. TOP: *Generator compartment with lights and electric cable reels.* COURTESY SAULSBURY APPARATUS COMPANY

Figure 3.5. BOTTOM LEFT: *Apparatus with light tower.* PHOTO BY AUTHOR

Figure 3.6. RIGHT: *Hydraulic extrication equipment compartment.* PHOTO BY AUTHOR

POWER UNIT COMPARTMENT

These compartments provide space for built-in power units or power units mounted on handcarts (Figure 3.6). The side walls of these compartments are outfitted with mounting brackets for hoses, rams, chains, slings, extrication accessories, or specially designed tools and equipment.

POWER SUPPLY COMPARTMENT

These specially designed compartments have reel-mounted hoses that make additional hoses quickly accessible.

AIR BAG COMPARTMENT

These compartments can be mounted to the underside of the flooring or bottom of any compartment. Configurations to accommodate air bags of different sizes and shapes can be specified so that the entire set of air bags can be stored in one compartment.

CASCADE SYSTEMS, COMPRESSORS, AND SPARE-CYLINDER COMPARTMENT

This storage area, common to many new apparatus, was designed to accommodate specialized equipment. Combining these features into part of a rescue apparatus is cost-efficient; it eliminates the need for a special apparatus for storing and refilling cylinders.

SAW COMPARTMENT

All power, chain, electric, circular, sawsalls, and handsaws are carried in this area.

TORCH AND BURNING EQUIPMENT COMPARTMENT

Large burning outfits, portable hand-carried or backpack burning kits, fire blankets, gloves, wrenches, spare tips, and protective goggles can be stored in this area.

ROPE AND RIGGING COMPARTMENT

Various sizes of ropes, rigging equipment, block and tackle, and hauling systems are kept in this compartment, which must be in an area that protects the rope from being damaged by the elements.

METERS AND TESTING EQUIPMENT COMPARTMENT

Oxygen, carbon monoxide, and combustible gas detectors; explosimeters; heat detectors; and similar types of items can be carried in this area.

EXTINGUISHER COMPARTMENT

Various types and sizes of extinguishing appliances and powders will fit in this area.

COLLAPSE EQUIPMENT COMPARTMENT

This compartment houses jacks, chocking, and shoring equipment; wedges; shovels; hammers; nails; tapes; buckets; and similar equipment.

SCUBA EQUIPMENT COMPARTMENT

Wet suits, dry suits, tanks, regulators, gauges, and other related items can be kept here.

HAZ-MAT COMPARTMENT

This area holds encapsulating suits, masks, in-line air systems, meters, and detectors.

FIRST-AID EQUIPMENT COMPARTMENT

Among the items usually carried are a first-aid bag, a trauma kit, burn kits, stretchers, splints, immobilizing kits, resuscitators, and long and short boards.

The equipment carried for primary types of responses usually dictates the designs for individual compartments (Figure 3.7). Adjustable shelving and slide-out trays can provide flexibility, but the problems a department experiences most frequently might indicate the need for a specially designed compartment and layout for the apparatus. A department that responds to many water-related incidents, for example, would want life-saving equipment together with first-aid equipment so that access is available to anyone assisting in the rescue operation. The time saved looking for the needed equipment could be time needed to save a life.

Figure 3.7. Compartments can be designed and constructed for specific needs. PHOTO BY
AUTHOR

New, customized, or specialized apparatus is not purchased every
day, and it is not built or delivered in a day. Therefore, those who are
going to have to work with it should be given time to offer the suggestions, ideas, and input that will provide the department with the most
efficient and functional apparatus.

Fire apparatus is unique and can be compared to a firefighter. Both
are often awakened in the middle of the night, started quickly, and out
the door without a warm-up; both race to the scene, expend lots of
energy in a short period of time, and then go back to the firehouse to
await the next "quick start." This type of life takes its toll on apparatus.
As much thought should go into choosing rescue company apparatus
as usually goes into hiring its firefighters.

In addition, the rescue unit's specialized tools and equipment, so
vital for its success, also must be appropriately suited to the unit's needs,
and its members' input should be a consideration when choosing the
tools and equipment to be used.

4

TOOLS AND EQUIPMENT

COURTESY J. CALDERONE

COURTESY J. SKELSON

CHOOSING SPECIALIZED EQUIPMENT and tools for rescue companies requires more than a Sunday flea-market attitude. The number of manufacturers producing these items has increased the department's options for selecting the equipment best suited to its needs. When purchasing tools and equipment for rescue operations, however, priority always should be given to product reliability, not price. Equipment must perform without fault when it's called for at a rescue operation. The fireground or rescue scene isn't the place for trial and error. A quality product kept in good working condition and operated by trained personnel will always be the key to successful rescue operations.

When a department needs a special tool or piece of equipment, who should do the shopping—the people who control the purse strings or the people who daily operate and maintain the equipment? Buying a special tool or piece of equipment without soliciting input or advice from those who use it is like buying waterfront property in the Sahara Desert.

Using members' expertise when choosing tools encourages innovation, which sometimes can make the difference between having or not having that specialized tool that can speed up the rescue operation.

An example of this is a rescue operation in which I participated on a hot July day when we used a piece of innovative equipment that a veteran and highly talented rescue firefighter recommended be incorpo-

45

Figure 4.1. *A-frame is placed into position using the apparatus winch. Large sections of concrete were removed from the basement area.* COURTESY CITY OF NEW YORK FIRE DEPARTMENT

rated in a new apparatus. Some of the brothers doubted the recommendation. The doubts later were erased, however, when the recommended A-frame assembly and winch were used to bring the incident to a successful conclusion (Figure 4.1).

A young girl was trapped under tons of concrete, the result of a sudden stoop collapse. To reach her, we had to remove much of the concrete. All of the surface debris, except a massive piece of concrete—the lone obstacle—remained. We backed a rescue apparatus positioned near the building into the operation's front area, connected an A-frame assembly, used a winch in conjunction with it, and successfully removed the large piece of concrete. Without that tool, the alternative would have been to remove all of the emergency apparatus that filled this small and narrow city street to bring a crane to the scene. The availability of the A-frame assembly and winch prevented a time-consuming operation.

Planning for purchases can save time and money and ensure getting the most out of the product. The department's needs must be considered first. Is the equipment to be used for only one type of operation, or would a tool that can perform that operation and more be considered? The extra cost for the multifunction tool could represent an overall savings, but only if it fulfills the department's needs.

If cost isn't a problem, the department is lucky. Larger departments have an advantage over the smaller ones: Buying in quantity can save a considerable amount of money. These departments should consider buying 25 spreading devices rather than one device if circumstances permit.

Choosing the Tools

Getting past the hurdle of price, what about the product's performance record? Should a department buy a product on its name only? Many older, established manufacturers benefit from their names because they have a proven product. Some departments seek advice from other departments who use brand "X" and purchase that brand based only on the recommendation. Other departments observe the tool in operation at an incident and become convinced that it's the one for them.

EXHIBITIONS AND SHOWS

How does one find out if the smaller manufacturer or "new kid on the block" has a product that's equally as good as an established product in price and performance? Manufacturers' exhibitions and shows offer the shopper a chance to view, inspect, and operate the equipment or tool. These events put the smaller manufacturer on an equal footing with the big one because the tool or equipment is available for first-hand inspection and demonstration.

Some manufacturers offer private demonstrations for departments that are not easily convinced or are unable to attend the mass demonstration (Figure 4.2); specific questions can be addressed and, more importantly, hands-on experience can be acquired at these demonstrations. Some manufacturers will give demos to small departments, and

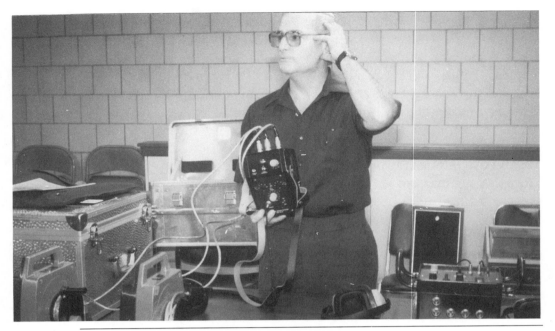

Figure 4.2. *Some manufacturers provide demonstrations for individual departments.* PHOTO BY AUTHOR

others will give them only to large departments. Factors that determine whether a company will give private demonstrations include the size of the manufacturing company, the size of the department, and the size of the potential order involved. In any event, most manufacturers will send literature on their products and usually follow up by phone.

VIDEO CASSETTES

A smaller company may not be able to afford to go to shows or exhibits, which can be expensive especially if the department's members must travel a great distance. For these companies, there is another way to get to know the tools and equipment: Enter the video cassette. In the fire service, tapes have been used for assessment and promotional exams, home-study programs, training programs, courses, schools, daily-drill subjects, and operational critiques. These demonstration tapes are unique promotional ideas. They provide the sales pitch; describe the tool's important specs, uses, and maintenance programs; and show its operating capabilities. The sales tape is becoming increasingly common and is an indispensable addition to the options the fire service can use to gather information for making purchasing decisions and vital improvements.

SEMINARS AND TRAINING EXERCISES

Another means to demonstrate tools and equipment is through seminars and training exercises. Vendors lend their tools and equipment to the organization running the exercise to use and evaluate during the training session (Figure 4.3). Some manufacturers offer tools and equipment to departments on loan, giving firefighters the chance to use them during rescue incidents. Being able to evaluate the tool under working conditions is an obvious advantage for the department, but it can work against the manufacturer. My department, for example, has a program under which tools and equipment are tested in the field for six months, and at least 50 percent of them have been found to be unsuitable for our needs.

Many years ago, a rescue unit was involved in the pilot testing of a power saw being introduced to the fire service. The saw's primary use was to vent roofs; it received some heavy action during its short stay. While cleaning the tool and getting it ready for the next job, we noticed that much of the fuel had to be replaced. Fuel was required even after short operations. A fault in the design caused the fuel cap to loosen and

Figure 4.3. Some vendors lend their tools and equipment to organizations running training exercises. PHOTO BY AUTHOR

the fuel to spill out during the saw's operation. The manufacturer redesigned the defective part, thereby eliminating this unit's potential for causing a disaster, and it went on to become one of the largest suppliers of saws for the fire industry.

Storage Alternatives

To be prepared for any type of rescue operation, most rescue companies would like to have as much equipment and tools at the scene as possible. Space limitations, department budget allowances, unit response assignments, and equipment presently assigned, however, are among the factors that dictate what's carried aboard a rescue company rig (Figure 4.4). A rescue company, for example, that's also a scuba unit must provide sufficient space for tanks, suits, weight belts, and other dive equipment. A unit providing hazardous-materials service needs additional room for the specialized suits; extinguishing agents; and leaking, sealing, and detecting equipment. Units that supply additional lights or salvage work require the necessary room for their equipment. Often, rescue companies add these services to the list of duties that they're capable of performing.

Figure 4.4. *Various limitations dictate how much equipment and tools rescue companies carry.* COURTESY J. SKELSON

"Housecleaning" is another important aspect of the tools and equipment detail. Obsolete equipment or equipment that's been superseded by a new and better model often must be removed from the apparatus. Although newer apparatus is designed to provide additional space, the space, after a few years, seems to disappear; it usually is filled with more equipment.

One solution for departments facing a lack of space is to stockpile equipment or tools that no longer can be carried on the apparatus—but which still can serve a useful purpose—at the unit's quarters or a central location; it also could be placed on spare apparatus, or a platform on demand (P.O.D.) Having the equipment on a spare apparatus or P.O.D. could be the ideal situation, as the equipment would be ready to go and have a means for bringing it to the scene. The spare apparatus also could replace an in-service apparatus should it break down or be taken out of service for repairs. Backup equipment should be stored aboard a spare apparatus or a P.O.D. in case the original equipment breaks down. Some departments have trained associated companies to perform the operational procedures should the extra apparatus and equipment be needed. Giving these companies a list of the tools and equipment will make it much easier for them, as they generally are not familiar with the specialized equipment.

A central location could be used for storing a whole host of additional equipment: shoring, cribbing, chocks, support jacks, foam, extinguishing agents, absorbent products, recovery drums, generators, smoke ejectors, and lighting equipment. Many major emergencies often require this additional equipment, especially during extended operations.

If equipment is to be stockpiled in the unit's quarters or at a central location, a separate and specially designed area should be set aside. Placing this spare equipment on an arrangement of numbered shelving makes it easy to prepare a list that indicates the type of equipment and its location (Figure 4.5). A copy of this list should be at the site location, and one should be given to the associated company. The list at the site location would be invaluable should a company other than the trained associated company be required to transport the additional equipment. Departments that have more than one central location should standardize their spare equipment and SOPs.

Some departments now are using companies as support units. These units are given some specialized equipment, which they are trained to use, and can provide assistance and equipment to rescue companies. Sometimes, these units will be at the scene prior to the rescue company, and they will have the capabilities to start setting up equipment while waiting for the rescue company to arrive.

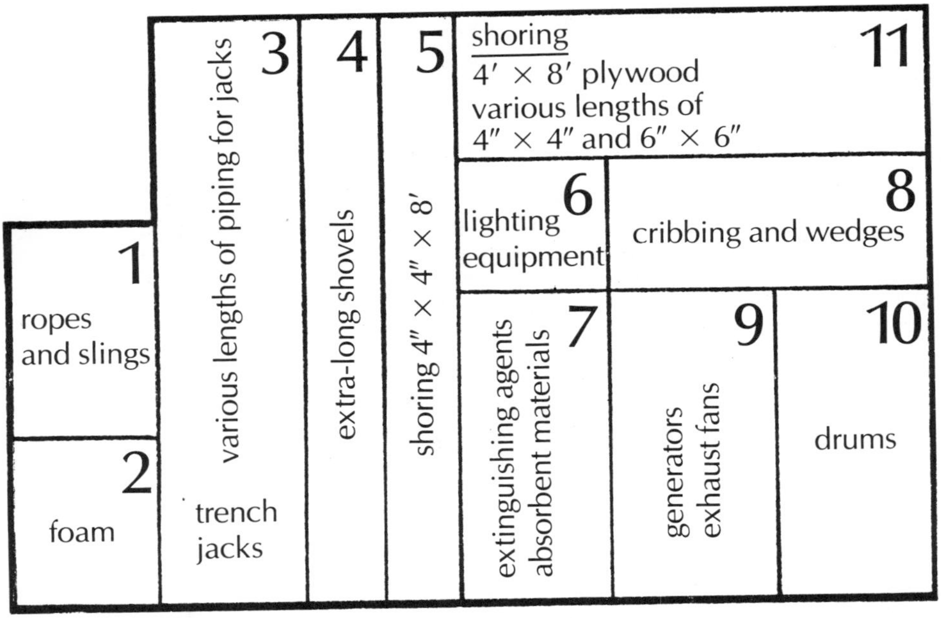

Figure 4.5. Spare Equipment List.

The Tool List

A tool list, mandatory in all rescue companies, serves many purposes. Many departments require that inventory listings of all tools and equipment be prepared on an annual, semiannual, or other specified basis. Understandably, the larger the department, the greater the need for control. This list provides the unit with an inventory record and a checklist for daily operation.

The list of tools and equipment can be arranged alphabetically. Each item should have its location on the apparatus listed alongside it. Compartments, which can be marked numerically or alphabetically, also can be noted on the tool list (Figure 4.6). This list makes it much easier, especially for those unfamiliar with the equipment and apparatus, to locate a tool or piece of equipment. A list of compartments and the tools and equipment contained within each compartment also can be of great help and can be placed in a waterproof plastic holder and attached to the inside of the door. Wall space can be used if the doors are of the roll-up design.

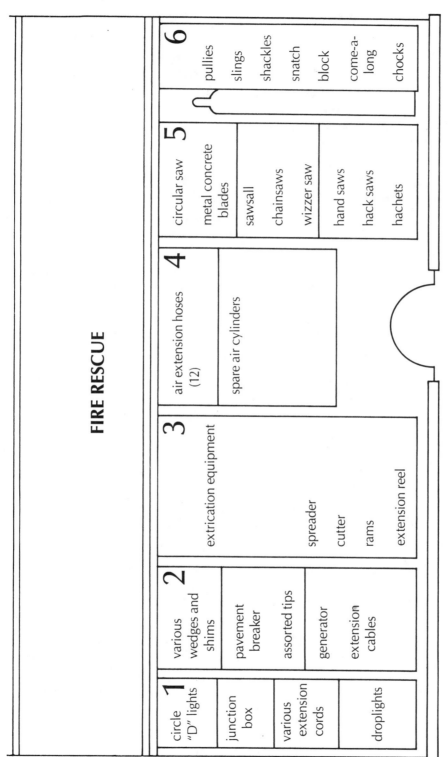

Typical schematic layout of tools and equipment carried on the rescue vehicle.
Similar charts should be prepared for all sides (inner and outer) of the apparatus, including the cab.

Figure 4.6. Schematic of tool layout in vehicle.

Schematic drawings of the various available compartment spaces often are useful for training purposes. I've often heard newcomers remark, "I can't believe there's that much equipment on the apparatus!" or "How do you remember where everything is located?" A complete set of the drawings should be kept with the apparatus, and one should be filed in the office company records. These drawings should show each compartment with the equipment listed in the appropriate space. Blank compartment sheets are most helpful for training new members. They can use them to practice remembering the various pieces of equipment in each compartment; the company officer can use them to test the member's knowledge. (Testing in this manner should be done only for positive reinforcement and not be used in a punitive way.)

Figure 4.7. Alphabetical tool list.

Tool	Compartment
Acetylene Large	3 b
Acetylene Packo	3 b
Air, 300 cu.ft.	11
Ambu Bag	13
Asbestos Blankets	3 b
Axes	Int rear
Amfire Drill	1
Back boards	Int UR
Back plates, Scott	4 a
Bailing Hooks	14
Bam Bam Tool	15
Beam Clamp	4 c
Bits, Pavement Bkr	9 c
Blankets	15
Blocks, wooden	2 b/c
Block & Tackle	10 b
Halligan Tools	Int/rear/16
Halligan Hooks	Int/Side bench
Hand Saw	8 b
Handle for Tirfor	6 a
Hard Hats	12
Hatchet	16
High-Pressure Hydrant Wrench	7

The three types of lists—alphabetical (Figure 4.7), compartmental, and schematic—all can provide additional help for members, especially those unfamiliar with a particular apparatus or piece of equipment. A binder containing the lists can be carried on the apparatus and be made available when needed. The list also can be very useful when taking up after a major operation where numerous tools and equipment are removed from the rig. The checklist provides an accurate accounting and reference for returning the equipment to their proper places. For departments with more than one rescue company, the ideal solution is to have identical apparatus, equipment, and tools with matching compartment design and storage layout.

The department's needs are the prime considerations when selecting tools and equipment, and their selections should be based on the input and advice of the members who operate and use the tools and equipment, remembering that rescue units are special. They require reliable tools and equipment that must be operated by highly trained firefighters.

Equipment Varies With Unit

When we think of rescue tools and equipment, we usually envision the large hydraulic spreading and cutting devices most commonly seen at a rescue operation. What tools are actually carried by the rescue companies throughout the country?

A number of factors determine the type, size, brand name, and amount of equipment each department carries. Some of them, such as apparatus size, compartment availability, and, the most important, budget considerations for special units within a fire department, are discussed in other chapters. Older, more established companies, of course, may have a stockpile of a variety of tools and equipment based on their years of experience in different types of rescue operations. Newer companies usually begin by acquiring some of the most common and standard tools and equipment carried by other rescue companies. Their inventory increases with time and experience. Budget restraints can have a very serious effect in this area.

The size of the inventory and the amount of equipment are not necessarily the answers to solving every rescue situation (Figure 4.8). As a young firefighter in a rescue company, my first introduction to extrication involved the use of the main tool at that time, the porto-power. Small in size, it was capable of spreading, extending, pulling, and lifting. Comparing it with the electric, hydraulic, or pneumatic equipment on the market today, I wonder how we accomplished as much as we did

Figure 4.8. The size of the inventory and the amount of equipment are not necessarily the answers to resolving every rescue situation. COURTESY J. SKELSON

with the equipment we had at that time. Yet, surprisingly, that piece of equipment still is being used in rescue operations because of its smaller size, its capabilities, and its power. Even though newer equipment may be available, the tool that works best for the situation should be used.

Ingenuity a Factor

Two incidents involving tools and equipment help to remind me that we must use not only the equipment available, but also the ingenuity of our talented rescue firefighters. The first incident involved a worker trapped in a building collapse. After working for more than two hours to free him, one obstacle stood in the way of the rescue workers. A small lifting device had to be placed in a narrow opening so that it could provide the final force needed to free the victim's foot. As the device was being placed, it was stopped by what appeared to be a piece of wood lath that the rescuers could not see. The smallest hand saw could not reach the area that needed to be cleared. The alternative was to remove a large amount of extremely unstable debris. One rescuer, feeling the wood and realizing its position and location, came up with the solution.

Reaching into his pocket, he opened a small multiblade knife. One of the blades was a saw no more than three inches long. Repositioning himself, he was able to remove the wood lath by using this small blade in a very narrow opening. Many of the rescue firefighters carry these small multipurpose knives. At times, ingenuity is more powerful than the largest tool or piece of equipment.

A new pair of binoculars came in very handy during a heat wave in a major city. The waterways in the area became the "backyard pool" for many of the inner-city residents. The fire department was receiving as many as half a dozen calls a day for "people in the water." This message prompts a response not only of the local companies, but also of a special Scuba team and a marine unit, in addition to a number of other city agencies.

One evening, as dusk was setting in, we received a report of a "victim" in trouble in the water. The waterway is known for its unusually rapid current, moving at a rate that presents problems for boaters. The fast movement of water and the dusk lighting made it appear that a "victim" was in great danger. The officer who first was notified found it unusual that although the "victim" was only 200 yards away, no screams or calls for help could be heard. Deciding to take a good look prior to calling for assistance, he put the new binoculars to use. What at first appeared to be an unconscious, life-jacketed victim really was a pile of debris someone evidently had thrown into the water. An unnecessary response and possible dive operation was averted due to the use of a new piece of equipment. I'm sure every rescue unit has had an incident in which the most unsuspecting piece of equipment or tool played a vital role in the outcome.

Power Sources

When choosing tools and equipment for the rescue operation, their power source must be considered: electricity—whether portable (two-cycle and four-cycle) generators (Figure 4.9) or standard power from utilities (AC or DC), pneumatic, hydraulic, hand-foot pump, rechargeable battery, gas, or other fuel. Most modern equipment is capable of being powered by more than one means. Hydraulic tools, for instance, can be supplied by generator power, electric power, and hand-operated pumps.

From the standpoint of safety, *alternate power sources* must be available at every rescue operation. For example, fuel and oil spills are characteristic of auto accidents, and a power supply for tools used at these operations must be chosen carefully to prevent ignition of the spilled

Figure 4.9. *Portable generator supplies electric power for lighting.* PHOTO BY AUTHOR

Figure 4.10. *Apparatus hose reels supply hydraulic extrication equipment with power.* PHOTO BY AUTHOR

material. Many rechargeable battery-operated and gas-operated tools common in the construction industry are finding their way into the fire service.

Application of equipment also demands the availability of alternate power sources. Extrication equipment can be supplied with power directly from the apparatus with hose reels capable of reaching areas within reasonable distances (Figure 4.10). But what if the tool or equipment is required at a distance or area not within the scope of the apparatus? The alternate power source (generator and hose reel) can be brought to a scene regardless of the distance because of its size and portable design. Information about the tools and their alternate power supplies should be a part of every game plan.

Component Purchasing

Smaller departments, those just starting out that may lack funding, might want to consider component purchasing. If the department cannot afford to purchase the complete set of equipment or tools, it should consider starter sets. Air bags, for example, can be bought in complete sets (10 bags of various sizes and lifting capabilities) or one at a time, complete with regulator and hoses. Additional bags then could be purchased as funding becomes available.

Tools Commonly Carried

The following is a list of equipment and tools commonly carried by rescue companies in the fire service:

EXTRICATION EQUIPMENT

These tools provide spreading, pulling, lifting, cutting, and extending capabilities. A wide range of these products are on the market. Their various sizes and weights provide options for rescue units with varying needs.

CUTTERS

These devices have different opening sizes and cutting capabilities. A combination *cutter-spreader* advertised as a one-person tool is available. It may be a good choice for a unit that lacks funds for a complete set of tools or that has space limitations.

RAMS

They are used to push or extend and come in various lengths and styles. Many models have removable heads that can be changed to fit the operation. The telescoping design allows two rams to be combined into one ram. Adapter kits that include chains, hooks, tips of different sizes and designs, and an assortment of parts increase the versatility of these units.

POWER SOURCES

They may be directly from the apparatus, from fuel or electric generators, or by hand/foot pumps. The foot pumps have a pedestal and the hand pumps, a grip handle for manual operation.

MANIFOLDS OR BLOCKS

They allow more than one tool at a time to be operated from one power source. Some manufacturers provide blocks, while others provide additional hydraulic outlets on the main power supply. Hydraulic line reels can be mounted on the apparatus; on the pump itself; or on portable, hand-carried assemblies.

The types, models, sizes, and selection of tools clearly are choices of the department and are to be based on its needs. Although often referred to as extrication tools, their capabilities surely are not limited.

Air Bags

When air bags were first introduced to the fire service, there was some confusion about their intended purposes, and they caused alarm among those unfamiliar with their capabilities. Used for many years in construction and other industries, their potential in fire service incidents has been proven.

Air bags are available in high- and low-pressure designs. High-pressure bags (Figure 4.11) generally operate up to 120 psi. Their lifting capabilities range from three inches to 20 inches in height and from one and one-half to 75 tons in weight. Depending on the sizes of the bags, it takes as little as one cubic foot or as much as 47 cubic feet of air to inflate them. The newer, smaller bags are ideal for lifting in locations of limited access where hand tools or other equipment prove inadequate.

Regulators, controls, hoses, and adapters increase their versatility and provide the means by which these bags can be used for a variety of

Figure 4.11. Air bags (high-pressure) are available in 10 sizes. PHOTO BY AUTHOR

Figure 4.12. Low-pressure bags are ideal for lifting heavy loads, especially on uneven, soft terrain or on ice- and snow-covered ground. PHOTO BY AUTHOR

rescue operations. A version of the bags was designed especially for hazardous-materials teams. Bags are available also for the following applications: sealing; patching; and stopping leaks in cylinders, tanks, tank cars, and trucks.

Bags have even been designed for use inside pipes, drains, and sewers, and provide a sealing condition from the inside. Their wide range of types and sizes make them readily adaptable to almost any need. Leak-sealing bandages for covering pipes are also available; their sizes fit smaller pipes.

Low-pressure bags are ideal for lifting heavy loads, especially on uneven, soft terrain or on ice- and snow-covered grounds (Figure 4.12). They're generally used in pairs for safety. Using as little as seven psi, 18 tons can be lifted up to a height of six feet. Low-pressure bags often are used for uprighting overturned tank trucks, and they're used quite extensively in outside industry.

Other Equipment

POWER SAWS

A variety with wood-, metal-, and masonry-cutting capabilities are part of the rescue company's ensemble: chain saws (electrically, hydraulically, gasoline-, and air-operated), circular saws, sawsalls, the Wizzer saw, and handsaws, including hacksaws, keyhole saws, and other types (Figure 4.13).

TORCH AND BURNING EQUIPMENT

Available in standard oxyacetylene setups, some larger outfits can be carried on hand trucks and by portable means (both backpack and hand-carried). Other portable units supply oxygen and use electrical power with a special cutting rod to provide a unique rescue cutting system. Protective equipment such as blankets, gloves, and safety goggles must be part of the system (Figure 4.14).

SMALL HAND-HELD AIR HAMMERS/CHISELS

They are equipped with various cutters (chisel bits, panel cutters, and bull points capable of cutting a variety of materials. In addition to these capabilities, the larger models can be used to breach concrete and masonry materials.

Figure 4.13. Power, electric, chain, and hand-held saws. PHOTO BY AUTHOR

Figure 4.14. Torches and burning equipment. PHOTO BY AUTHOR

Figure 4.15. Electrically operated "demo" hammer.

Figure 4.16. Meters and detectors. PHOTOS BY AUTHOR

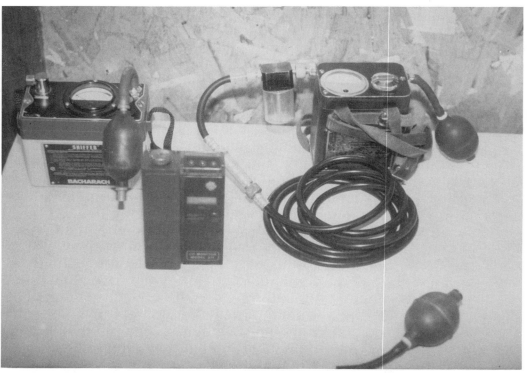

PAVEMENT BREAKER, JACKHAMMERS

They are pneumatically, hydraulically, electrically, or fuel-operated and provide cutting, drilling, and breaching of heavy concrete or asphalt (Figure 4.15).

METERS/DETECTORS

Those found aboard the rescue truck might include explosive meters, oxygen indicators, carbon monoxide meters, gastek meters, electrical meters, heat detectors, thermal cameras, radiation detectors, and various chemical test kits for chlorine, hydrogen sulfide, and sulphur dioxide (Figure 4.16).

LIGHTING EQUIPMENT

Included are portable lights, reels, fittings, converters, junction boxes, and adapters. Some apparatus lighting is provided by air-operated telescoping extension lights permanently attached to the apparatus and capable of rotating 360 degrees in all directions. Other equipment is removable and includes portable telescoping light mounts that can be powered by a generator or electrical extension lines connected to the apparatus and capable of being moved from location to location.

New items include small, portable generators with high-powered lights. Their size allows one person to move them easily to different locations. Much of the new equipment is waterproof, crushproof, and engineered with firefighter safety as a priority. The amount and type of lighting equipment needed depend on the unit's response duties, as directed by its department policies. Some departments use specialized apparatus that provides lighting with other capabilities, such as air-cascade systems.

ROPE AND RIGGING

On the rescue apparatus, this category includes various types of rescue ropes, slings, hooks, shackles, snatch blocks, harnesses, ascenders, carabiners, figure eights, rappelling rigs, special rescue slings and harness, and rescue litter and stretcher systems.

COLLAPSE EQUIPMENT

Included might be trench jacks (Figure 4.17), air-shore jacks, cribbing, chocks, shims, shovels, pry bars, assorted timbers, planking,

Figure 4.17. *Trench jacks used for support during collapse incidents are carried by rescue units.* COURTESY CITY OF NEW YORK FIRE DEPARTMENT

sledgehammers, wrecking bars, pinch bars, measuring tapes, hammers, and nails.

SELF-CONTAINED BREATHING APPARATUS

They should be supported by additional cylinders, large-capacity air bottles, air-supply lines, and supply systems for use in confined-space and other rescue applications.

MECHANIC TOOL SET

This set includes pipe wrenches, pipe cutters, bolt cutters, utility shut-off keys, plug kits, wedges, sealants, fittings, and couplings.

FIRST-AID EQUIPMENT

That carried aboard the rescue rig includes a first-aid bag, a trauma kit, burn kits, stretchers, splints, immobilizing kits, resuscitators, and long and short boards.

REMOTE PHONE AND HANDIE-TALKIE SYSTEMS

A variety of these systems should be part of the special sound-powered phone equipment (Figure 4.18) used by the rescue company.

Units with special duties such as scuba responses, haz-mat responses, foam, salvage, or lighting in addition to their rescue duties carry that specialized equipment along with standard rescue equipment.

This list gives some idea of what is being carried by rescue companies; however, many units carry more and some, less. The amount carried generally is limited by apparatus and compartment space availability; imagination; experience; and, of course, budget restrictions.

Regardless of the amount, type, and size of the equipment and tools assigned to a rescue unit, all must be maintained in good operating condition and be capable of performing up to standard when called on in rescue operations.

The rescue company's tool and equipment inventory does not remain constant. As already noted, it must be evaluated regularly and adjusted to meet the department's changing needs and to reflect a constantly evolving technology. This new knowledge renders some tools and equipment obsolete and delivers in their places state-of-the-art versions that further enhance the rescue company's capabilities.

Figure 4.18. Sound-powered phones, headsets, and reels. PHOTO BY AUTHOR

5

STATE-OF-THE-ART EQUIPMENT

REPRESENTATIVES OF FIRE DEPARTMENTS throughout the world gather together at conferences, conventions, and symposiums to discuss common issues and concerns. Lectures, demonstrations, and hands-on training give these attendees opportunities to discuss methods and strategies for dealing with the situations and problems they encounter. And, it is during these conferences and conventions that apparatus and tool/equipment manufacturers, distributors, and vendors exhibit the state-of-the-art firefighting and rescue products. Many innovative changes in apparatus or tool/equipment design directly result from the information exchanges between users and manufacturers that take place at these events.

Multipurpose Apparatus

Tools and equipment are becoming more sophisticated, and yet they are more portable and adaptable. Each year the capability and the versatility of apparatus appear to grow. Some of today's apparatus combine a multitude of services into one special heavy rescue unit. One example is a combination rescue and haz-mat unit that can be used as an incident command station and features a fully equipped communications system. This type of combination unit gives the most for the dollar.

69

Figure 5.1. *Slide-out trays make access to tools and equipment easier and safer.* COURTESY SAULSBURY APPARATUS COMPANY

Figure 5.2. *Compartment with drop-down doors for storage of equipment.* PHOTO BY AUTHOR

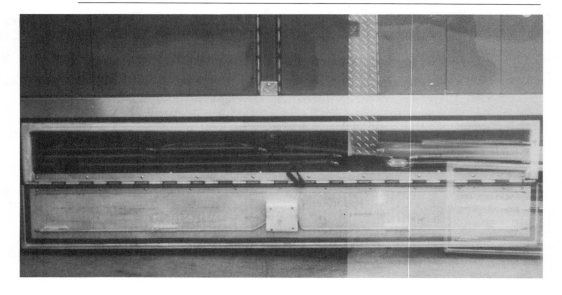

In addition, compartments in these new rescue apparatus are designed to accommodate the large extrication equipment. Slide-out trays and mounts enhance the user's speed and safety (Figure 5.1). Every inch of space is used in designs that specifically conform to the unit's needs. Compartments with drop-down doors for storing air bags, blocks, chocks, and cribbing are attached to the underside of the apparatus, increasing storage capabilities (Figure 5.2). High-performance, built-in generators provide the electrical power source for lighting and tools. Electrical cable reels, in compartments, provide hundreds of feet of cable for remote operations. Light towers, some as high as 40 feet, now are found in front and/or in the rear of the apparatus. As an option, these air-powered telescoping light towers can be recessed into compartments when they are not in use (Figure 5.3). Cable reels also are available for air operations; hundreds of feet of cable on reels enable a tool or piece of equipment to be used at distances away from the apparatus without the need to bring the air source to the location.

Yes, the rescue apparatus of the 1990s can provide communications, generators, lighting, an incident command station, a cascade system, compressors, cable reel for air and hydraulic tools, winches, A-frames, and enough compartment space to handle various rescue tools and equipment (Figure 5.4). Booster tanks, foam tanks, and pumping capabilities also can be incorporated into the apparatus.

Platform on Demand

A system gaining in popularity in the fire service is the Platform on Demand (P.O.D.). Basically a "container" used for storing special equipment, such as haz-mat, collapse, communications, or designed for medical responses, command stations, or a combination of several special units, P.O.D.s are stored at a central location and brought to the scene of an incident by a truck (Figure 5.5). Because of the versatility of the containers and the truck, unlimited containers can be centrally stored and brought to a scene using only the one truck.

Another apparatus on the market is a four-wheel drive suburban first-responder vehicle that has a full set of extrication equipment mounted on slide-out trays and additional space for other equipment. This vehicle would be ideal for departments that have trouble gaining access to areas in their response district with the larger, conventional apparatus. This type of vehicle can be designed for a department's special needs (incident command, haz-mat, and so on).

Figure 5.3. *Air-operated telescoping light tower.*
PHOTO BY AUTHOR

Figure 5.4. *Apparatus of the 1990s combine a number of features: The apparatus can be used as a heavy rescue unit, an incident command center, or a haz-mat unit.* COURTESY SAULSBURY APPARATUS COMPANY

Figure 5.5. *Platform on demand (P.O.D.) and tractor used for transporting.* PHOTO BY AUTHOR

Self-Contained Breathing Apparatus

Self-contained breathing apparatus constantly is being modified and redesigned to meet fire service needs while complying with federal regulations and recognized national standards. The newer SCBAs are constructed of lightweight material that is stronger and more durable and can better withstand abuse during firefighting operations (Figure 5.6). With the improvements in design and modification have come such features as a rescue mask—a second mask carried by the user that can be

---

Figure 5.6. *State-of-the-art SCBAs are constructed of lightweight material that is stronger and more durable than that used in older models.* PHOTO BY AUTHOR

connected to an SCBA's quick-connect valve to supply a victim with air. A buddy breathing system uses the same type of quick-connect coupling to enable a user to supply air to another firefighter via a short-rescue hose that links together in seconds the two SCBAs. In-line extension systems with small, compact escape cylinders for use in confined-space operations eliminate the additional hazards of the bulky equipment normally used.

Communicating while wearing SCBA always has been a problem. A number of systems now available allow the sender and receiver to talk and hear clearly without interference: in the mask radios, speech-activated diaphragms, wireless communications, and masks with push-to-talk buttons.

Personal Alert Safety Systems

A major concern in every rescue operation is the safety of the rescuer as well as that of the victim. Personal alert safety systems have been available for a number of years. Now available is a small (two-inch x two-inch), lightweight (three-ounce) personal motion-sensing alarm with a loud (96-decibel) audio alarm and a bright visual flashing indicator that activate should the wearer become motionless for 25 seconds or longer. The alarm easily can be attached to clothing, belts, SCBAs, or helmets.

Hydraulic Tools

Manufacturers of hydraulic rescue tools for vehicle extrication have introduced a number of variations to the spreaders and cutters that have been on the market since the early 1970s. A combination lightweight, compact tool that can spread, pull, or cut is ideal for departments whose budgets prohibit purchasing an entire system. Newly designed jaws provide a greater grabbing surface with less slippage and can be changed over in seconds. One manufacturer has replaced the chains used for pulling steering columns and displacing dashboards with heavy-duty straps that are much lighter and less cumbersome. Another system available is portable, lightweight, and air-operated with standard air bottles, air brake systems, or a cascade system. Hydraulic rescue tools now can be supplied from an apparatus complete with cable reels and power sources, entire systems mounted on wheeled hand trucks, or systems that are portable and light enough to be hand-carried to the scene.

Special Saws

Most power saws have the capability to cut wood, metal, and concrete simply by changing the blade. A new saw designed strictly for cutting concrete (Figure 5.7), block, reinforced concrete, natural stone, brick, and similar materials uses a 14-inch diamond blade on an off-center drive that gives an extreme cutting depth of 10 inches. Compared with a conventional center drive, this is quite an advantage. A hydraulic power pack that serves as the power source is mounted on a small, wheeled hand truck and rapidly can be set up and moved from one location to another.

Cutting Systems

Although conventional oxyacetylene-cutting systems still are standard for rescue units, a new cutting technology that complements these systems is available. Ultrathermic cutting systems comprise a hand-held torch; consumable alloy fuel cutting rods; an oxygen supply with regulator, cables and hoses; and a power source, which can be a 12-volt bat-

Figure 5.7. Hydraulically operated concrete-cutting chain saw. COURTESY J. NORMAN

Figure 5.8. *The latest in cutting systems, the plasmar cutter, uses high-voltage and air. It has all the features of other cutting systems, but it lacks portability.* PHOTO BY AUTHOR

tery or a built-in ignitor system. Varying in size, the ultrathermic systems are available on hand-truck-mounted systems, in carry-case models, and in back-packed-mounted systems for the ultimate in mobility. They differ from traditional oxyacetylene systems in that they use consumable alloy fuel rods that burn at temperatures in excess of 10,000 degrees F instead of at 5,800 degrees F, as acetylene does. Because of these high temperatures, materials liquefy almost instantly so that preheating is not required and cutting is more quickly accomplished. A variety of these systems is available; departments, therefore, may choose the system with the features that meet their individual needs (Figure 5.8).

Portable Power

Portable lighting and power sources are musts for rescue operations. Realizing that many locations make using an apparatus-supplied system impossible, portable power must be available as a backup. Small, lightweight, and compact generators that are easy to carry and operate and provide up to four hours or more of quiet operating time can replace those bulky, noisy, commonly used generators. These generators are available in a variety of sizes and power-output capabilities.

Detecting Devices

Finding that elusive source of heat and fire has been made easier (in some instances) by using heat-detecting or thermal-imaging cameras (Figure 5.9) that pick up the temperature differences and, in many instances, save invaluable time, energy, and equipment in isolating and extinguishing fire. A variety of this type equipment with price tags to meet department budgets is available. Detectors used for checking the presence of hazardous gases and vapors, carbon monoxide, or oxygen levels are manufactured in small, hand-held designs. These combination detectors are multifunctional and can replace the larger detectors used in the past. Many of these detectors use a LED display, which is particularly effective during night or dark interior operations. Operated by a

Figure 5.9. The thermal-imaging camera is used to locate hidden heat sources. PHOTO BY AUTHOR

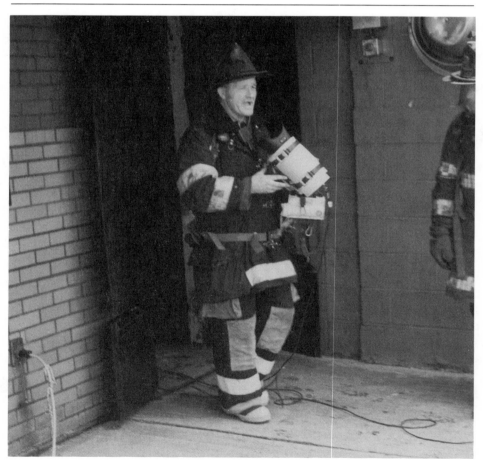

rechargeable battery, these detectors can be used up to eight hours before recharging becomes necessary. A special confined-space kit that includes a detector, charger, probes and hoses, earphones, and other accessories that provide pre-entry monitoring is an essential piece of equipment that should be carried by rescue units.

A popular hydraulically operated forcible entry tool has been retrofitted with interchangeable heads of different sizes, shapes, and capabilities, which effectively have increased its versatility.

The need to increase search-and-rescue capabilities has been magnified to a large extent by the devastating disasters—especially the earthquakes that occurred in Armenia, Mexico City, California, and the Philippines. The difficulties encountered while trying to locate the numerous trapped victims of these disasters have prompted the fire service and the manufacturers and designers of special technological equipment and devices that could provide valuable assistance to rescuers to develop equipment that will be effective in these incidents.

A number of "life-detecting" systems are now on the market; they continually are being tested and upgraded to improve the means by which rescuers can locate victims trapped under rubble (Figure 5.10). Using sensors, either seismic or acoustic, or, in some equipment, a combination of both, this equipment detects the sounds or vibrations coming from within a rubble pile. A variety of systems is used to monitor the signals; these signals are received through headphones and are displayed on meters which then are coordinated to pinpoint the victim's location on machines with LED readouts or to print out on display paper. Data interpretation then is used to coordinate possible victim location. Some of these designs are similar to the types used in mine safety and by the military to detect and locate movements of personnel or vehicles. They vary in physical dimensions (weight and size); one system is designed to be worn on the person and can be operated and carried by one rescuer.

As the advancements in modern technology increase, we can expect to see the present day victim-locating devices refined to the degree that they will be extremely effective for rescuers and victims alike. It often has been asked, "What if the victims are unconscious or unable to produce sounds or vibrations that can be picked up by these seismic or acoustic devices?" Again, modern technology and its adaptability for the fire service has provided us with fiberoptic borescopes and fiberscopes. By penetrating small access openings in rubble or by using holes drilled by rescuers, the flexible fiberscope or rigid borescope can be inserted and the interior inspected for trapped victims. The scopes have built-in lighting that enables rescuers to effectively view the inspection area up

Figure 5.10. Detecting systems are used to locate victims trapped under rubble. PHOTO BY AUTHOR

Figure 5.11. "Rescue sled" can be paddled in water, towed on snow, or slid on ice. COURTESY J. SKELSON

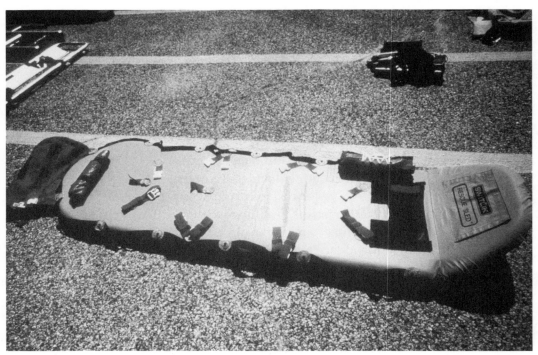

to a distance of 70 feet. The inspection area can be viewed on special TV monitors. This equipment now is part of the U.S.A. International Disaster Response Team and many rescue team inventories.

Transporting Devices

Devices for transporting victims are available in many designs. Besides the lightweight, breakaway stretchers and baskets, a new inflatable multisurface sled that can be paddled, towed, or slid on water, snow, or ice is being included in most rescue unit inventories (Figure 5.11). It comes with restraint straps, towing attachments, hand grasps, and other unique features.

The Space Age and Computer Age have had a direct impact on and have greatly influenced many of the rescue tools/equipment being carried by rescue units in the 1990s. Departments that have not seen some of these state-of-the-art tools and equipment should attend a conference, convention, or symposium. While there, they should not hesitate to offer to the manufacturers or distributors any suggestions they may have regarding specialized equipment. The best equipment is that which has been modeled after the ideas submitted by those who use it in real-life situations.

Even the most costly equipment will break down if it is abused and neglected. Given the urgent nature of the rescue company's work and the crucial role that ready-to-use specialized equipment plays in successfully completing that work, the rescue company must make maintaining that equipment a priority.

6

MAINTENANCE OF TOOLS AND EQUIPMENT

PHOTO BY AUTHOR

WHEN THE OWNER OF an older vehicle is asked, "How did the car last so long?" the usual reply is, "I changed the oil every 2,000 miles, checked all belts and cables; I really took care of it." That reply is a roundabout way of saying that proper maintenance and care were given to the vehicle. There is a message in this exchange for the fire service: The apparatus on scene may be the finest available, and the rescuers may be the best trained, but the operation still could be a failure. Why? Because the tools or equipment were not properly maintained.

Daily chores in quarters usually require a routine that includes maintaining the facility and apparatus. The tour starts with cleaning the kitchen, sitting room, apparatus room, and company offices (that's after 10 cups of coffee); a good sweep and swab are the orders of the day. The apparatus chauffeur is busy checking the levels of fuel, oil, water, and transmission fluid of the apparatus. The tool man is checking hand tools, extinguishers, and equipment used on most routine runs. But what about the extrication equipment, saws, generators, lights, air-operated tools, torches and cutting equipment, meters, rigging equipment, scuba gear, haz-mat equipment, and salvage equipment? The list also must include the equipment stored in quarters, in special apparatus, or in P.O.D.s.

The less specialized equipment a department has, the easier it is to organize a maintenance schedule. A heavy-duty rescue company with

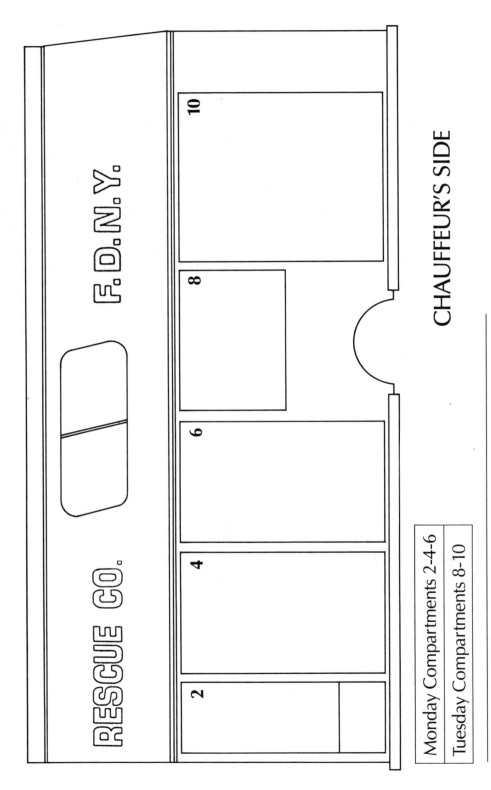

CHAUFFEUR'S SIDE

Monday Compartments 2-4-6
Tuesday Compartments 8-10

Figure 6.1. *Assigning work schedules for various shifts based on the compartment-numbering system ensures that maintenance will be systematic and complete and that chores will be evenly distributed among unit members.* PHOTO BY AUTHOR

more equipment and tools must expand its daily chores and work load to include a schedule that allows sufficient time to maintain its numerous tools and equipment properly. A true appreciation of how much equipment is carried on the apparatus comes with having to strip it for a changeover. If the unit's inventory includes equipment and tools that can be maintained and cared for on each tour, then each shift must follow the schedule.

Maintenance Scheduling Options

A larger inventory control system for maintaining and caring for equipment and tools can be handled in a variety of ways.

THE COMPARTMENT-NUMBERING SYSTEM

This approach facilitates systematic control because numbered compartments are assigned to various shifts (Figure 6.1). Schematic layouts of the numbered compartments also can be an excellent inventory-control mechanism. One option is to institute a weekly schedule that assigns the different shifts to care for specific compartments. Alternating compartments and shift assignments allows all personnel eventually to be responsible for the entire inventory over a specific time period and ensures that all equipment is properly maintained.

When using the numbered-compartment system, the schedule must be balanced. Some personnel may be assigned compartments that hold equipment needing only a visual inspection, while others may have to perform both visual and operational inspections.

Another way to divide the work schedule is to label the compartments according to the location on the apparatus (for example, "officer's side," "outside," "chauffeur's side," "inside"). The schedules should be set up so that the compartments located in a specific area are the responsibilities of the shift on duty.

The Grouping System

Another system that may be followed when establishing a maintenance schedule is to group the tools and equipment and to schedule maintenance duties for different groups on different days, as is shown in the following sample schedule (Figure 6.2):

MAINTENANCE SCHEDULE

Monday

COMMUNICATIONS EQUIPMENT
    A. RADIOS
    B. SOUND-POWERED
PHONES
PERSONAL SAFETY DEVICES

Tuesday

EXTRICATION EQUIPMENT
    SPREADERS
    CUTTERS
    RAMS
    EXTENSION REELS
    POWER SUPPLIES

Wednesday

LIGHTING EQUIPMENT
LIGHTS
GENERATORS
REELS
EXTENSION CABLES
JUNCTION BOXES, ADAPTERS

Thursday

METERS
TESTERS
EXPLOSIMETERS, CO TESTER
O_2 INDICATOR, ELECTRICAL TESTER
THERMAL CAMERA
RADIATION DETECTORS

Friday

AIR TOOLS;
    AIR HAMMER-CHISELS
    PAVEMENT BREAKER
    SAWS, DRILLS
    AIR BAGS, REGULATORS, ETC.

Saturday

SCBA
IN LINE EXTENSION HOSES
ROPE - LIFE SAVING
RIGGING EQUIPMENT

Sunday

SCUBA EQUIPMENT
HAZMAT EQUIPMENT
FIRST AID, TRAUMA KIT, BURN KIT
SALVAGE EQUIPMENT

Figure 6.2. Grouping tools and scheduling maintenance duties by the days of the week.

- *Monday*: All communications equipment (portable radios and sound-powered phones and reels) and personal safety devices
- *Tuesday*: Extrication equipment, including all components, cutters, spreaders, rams, extension reels, and power supplies
- *Wednesday*: Lighting equipment (lights, generators, reels, converters, adapters, extension cables, and junction boxes)
- *Thursday*: Meters and testers; explosimeters; carbon-monoxide tester; oxygen indicator; electrical testers; gastechs; thermal cameras; radiation detectors; and kits for detecting the presence of chlorine, hydrogen sulfide, and sulphur dioxide
- *Friday*: Air-operated equipment (air hammers/chisels, pavement breakers, saws, and drills), air bags, regulators, controllers, hoses, and accessories
- *Saturday*: Self-contained breathing apparatus with air-extension lines and systems and rope and rigging equipment
- *Sunday*: Specialized equipment for that particular unit (scuba, haz-mat, first-aid, trauma kit, burn kit, salvage equipment).

Operational procedures should require that all equipment and tools be checked after each use and before being placed back in service. The daily, weekly, or biweekly schedule must include an operational check in addition to a visual inspection (Figure 6.3). The following must be checked: fluid levels; hoses for cracks, breaks, splits, or kinks; connections, bolts, and screws for tightness; and general cleanliness, to ensure that all equipment is in top operating condition.

Figure 6.3. The daily, weekly, or biweekly schedule must include an operational check in addition to a visual inspection. PHOTO BY AUTHOR

The checklist should include the tool designation, manufacturer recommendations and guidelines for maintenance, special considerations gained from actual tool usage, the date of the inspection, and the signature of the inspecting firefighter. The checklist can be kept in a binder in the company office or in a special folder stored in the compartment with the tool or equipment. Comments noted on the checklist must be followed up and passed on to all shifts.

Scheduling maintenance checks and routinely taking care of equipment are beneficial in several ways. The maintenance schedule can be used for inventory control; it and the tool list can point out missing equipment or equipment that is out for repair but that had not been designated as such. A regularly scheduled visual and operational inspection also familiarizes rescuers with tools and equipment, helping them to use the tools with more confidence in an emergency operation.

Drills and training sessions used as part of the inspection routine generate interest. On a day when an air-bag maintenance inspection is scheduled, for example, a drill also should be planned to discuss and use operational procedures. It's important, however, not to get so caught up in the drill that the inspection and operational checks are forgotten (Figure 6.4).

Figure 6.4. Drill sessions are a good time for visual and operational checks. The required maintenance checks should not be eliminated, regardless of how involved the drills are. COURTESY PARATECH, INC.

Following a regular schedule also ensures that the tools and equipment used less frequently are checked and inspected. Again, it also helps to familiarize the rescuers with the equipment less frequently used.

Regardless of the amount or type of equipment the department's apparatus carries, regularly scheduled inspections and operational checks are musts. The scene of an operation is not the place or time to be wondering whether the equipment works. Properly maintaining and caring for the equipment is a responsibility that adds to the professionalism of the fire service.

Carefully chosen and properly maintained apparatus, tools, and equipment are vital components of a rescue company. Equally as important, however, is the expertise of the unit members who will be using these components. The availability and scope of training options for some departments also are being constrained by budgetary considerations. Innovative and resourceful departments, however, can provide adequate training for their rescue companies despite these restrictions.

7
TRAINING

COURTESY RESCUE COMPANY 2

THE COMPLEXITY AND DIVERSITY of specialized rescue unit operations necessitate training unit members on a continuous basis. Initial training provides the foundation; continuing training reinforces the principles of rescue operations and procedures. Departments should answer the following questions about their specialized training programs:

- What special training is offered?
- Who provides the training?
- Where is the training given?
- How are the department's training goals accomplished?
- Which subjects are covered?

The answers to these questions will vary according to the factors that influence individual department policies with regard to specialized training.

Firefighting is a unique profession; it takes a special kind of person to enter a burning structure when everyone else is exiting. It's training that enables the recruit to develop that strength and the energy to perform the required tasks safely. Training, therefore, is a top priority in most departments, and it is even more important in a rescue company.

Specialized Training

Many departments, for example, have training schools that cover a wide spectrum of rescue-training operations. The list is determined by a department's budget, imagination, dedication, commitment, and other resources.

These programs, some running for successive weeks, cover rescue tools, equipment, and operations. Their scope is influenced by factors such as the type of equipment carried by the units, the response of the units (rescue, EMS, haz-mat, scuba), the facilities, the availability of qualified instructors, and the special certification requirements imposed by government bodies. Some regulatory agencies require certification to handle extrication, haz-mat, and scuba operations. Many departments are capable of operating these schools on a small scale, but with high proficiency. In some cases, when a budget does not permit members to attend classes on a weekly basis, a unit is brought to the training site for a half- or full-day's training.

The training school (Figure 7.1) is an excellent place to introduce the new specialized equipment. One-day sessions could be used to train in such areas as extrication—new techniques and equipment—mask operations—confined space (Figure 7.2), manifold, extension-line use, and the like—and rope usage (Figure 7.3)—rescue, rappelling, stokes, or litter lifting.

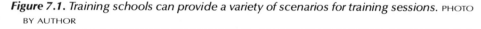

Figure 7.1. *Training schools can provide a variety of scenarios for training sessions.* PHOTO BY AUTHOR

Figure 7.2. *Confined space training using manhole at training academy.* PHOTO BY AUTHOR

Figure 7.3. *Specialized rope rescue training being conducted at training school.* PHOTO BY AUTHOR

Money usually is the main obstacle, especially for departments seeking to expand their training programs. When city management begins cutting the budget, the fire department often is the first to feel the impact, since fire department managers must make cuts and training is the one area the cost-cutters target.

National Programs

Members can receive specialized training also through courses offered by the national fire service organizations at the state, county, and sometimes even local levels.

The National Fire Academy (NFA), for example, since 1980 has been providing programs that are extremely helpful to any department. The courses are offered in a number of states as well as at the National Emergency Training Center in Emmitsburg, Maryland. *Fire/Emergency Services Sourcebook, Second Edition, 1990–1991* (Specialized Publication Services Inc., New York, 1989) lists all courses, the locations at which they are given, and directions for applying. The Fire Service Technical Specialist Program, which provides training in hazardous materials, is an example of an NFA program available at the state level. Another course teaches how to design, develop, and implement training courses/programs. Financial assistance from the federal government makes the NFA programs more accessible.

The Federal Emergency Management Agency (FEMA), the parent organization of the NFA, partially funds the state-run, rescue-training programs available in some areas. This certificated program covers victim rescue from entrapment in building collapse, operations in confined space, search and removal from mountainous areas, hoisting and lowering victims and personnel from above or below grade, rigging, gin-pole construction, tunneling, and trenching.

Many states sponsor programs directed to all departments within their borders. Maryland, New York, Oklahoma, Pennsylvania, Texas, and Virginia are just a few of the states that offer courses directly related to rescue; a more complete listing is given in *The Source Book*.

Towns, municipalities, counties, and other local jurisdictions also offer specialized training. A rural area with farms in its response jurisdiction, for example, would be interested in a course or program that offers training in handling farm incidents. One such course offered at a county training academy covers farm machinery, farm chemicals, and farm-structure incidents. The curriculum includes the extrication of manikins at simulated accident scenes.

Departments with budgets that do not allow for a one- or two-week course and do not have an NFA or similar program available can consider other options such as seminars, conferences, and workshops (Figure 7.4). Many are offered, and they cost less, run for a shorter period of time, and are held on a regular basis. Some of these smaller workshops, seminars, and conferences are offered on weekends to accommodate firefighters whose schedules do not allow them to attend during the workweek. Many states also offer weekend programs. Listings can be obtained from the state office, and many fire service publications list such upcoming events. If budget allocations still are a problem, a department can send only one member or two members to the training sessions, and they then can pass the information along to the other members of their units.

Private Organizations

A number of private organizations also offer courses with curricula that range from the basic to the advanced levels of rescue work (Figure 7.5). Some of the more advanced courses require prior experience or prerequisites based on the course program. Courses offered by private organizations should be checked to see if they are accredited by federal, state, or local government agencies.

Company Programs

Company-level programs are the heart of a unit's training. They generally are inexpensive and provide hands-on education geared directly to the unit's needs and objectives. The training office should schedule the training sessions, and schedules should be flexible enough to provide training for every unit member. Attendance at these sessions should be recorded and the records kept for future reference (Figure 7.6)

Effective scheduling hinges on accurate record keeping. A log of the tools and equipment used at each rescue operation, the members who used them, and an evaluation of the competence levels of the users are helpful to the training officer. (This kind of record keeping also is helpful when the department is considering the purchase of new equipment—justification for purchase can be supported by complete tool-use records.) The training officer may see the need for "refresher" training with equipment that's used on a limited basis; up-to-date records would be useful in this case also. The value of refresher training can't be overestimated; if the tool or piece of equipment is used only once and it saves

Figure 7.4. These members are participating in a rescue workshop. PHOTO BY AUTHOR

Figure 7.5. Private organizations offer advanced level courses, especially in the scuba field.
COURTESY L. SANCHEZ

TRAINING RECORD

FIREFIGHTER _____ DATE ASSIGNED _____

Tool/Equipment	Lecture	Drill	Officer/Comments
AIR BAGS	3/15	3/17	Capt. D - Good Drill
AIR TOOLS			
CO TESTER			
CHAIN SAW			
EXPLOSIMETER			
HURST TOOL			
OXY-ACETYLENE			
ROLLGLISS			
* ALL EQUIPMENT AND TOOLS THAT UNIT CARRIES SHOULD BE LISTED IN THIS COLUMN			

Figure 7.6. *Individual training record.*

a life, it's well worth its cost, and especially the cost of the time spent to train members in its use.

One point deserves to be reemphasized: Training must be on a continuous basis. Ideally, the longer the time allotted for training, the greater the variety of topics that can be covered. Week-long programs are better than shorter sessions for covering a greater number of tools and equipment, but it is not the quantity but the quality that counts. Having an instructor who can "show" but not "tell" is not much help for the rescue member who wants the "meat and potatoes" of the program. Planning for specialized training should include choosing a qualified instructor who has the background to give the rescue members the special instruction needed.

Keeping a separate file or folder for the *instruction books and manuals* of each tool and piece of equipment is another important aspect of rescue training. These files can be used for the "introductory phase" of the training, particularly for the new members. Manuals provided by the manufacturer explain tool operation and maintenance and include the parts list; but, more importantly, they show various examples of the tool's possible applications in rescue situations. Expanding a tool's capabilities is the result of training and practical application.

In addition to tool manuals and instruction booklets, other training aids that should be considered are *videotapes, slides, 16-mm films, articles* written about rescue operations, and manuals designed for training rescue personnel.

The *videotape* is a very effective training tool. Many of the major rescue equipment manufacturers can provide a tape describing the equipment's capabilities, limitations, maintenance requirements, and design features. A number of specialized training videotapes on the market provide step-by-step instruction for basic, difficult, and unusual operations. Tapes are being used in many programs across the country and have proved to be excellent training tools.

Many units are making their own tapes, recording their training sessions, and using them for "classroom" instruction at the firehouse. Critiquing the tape can improve operations and ensure that the safety of all personnel is being considered. Refinements of the operation can be retaped and kept on file for members' use at any time. Departments that do not have a video recorder or video camera may be able to borrow the equipment from its members.

Slides and 16-mm films are very effective for in-house productions of drills. With videotapes, they can provide an extensive training-aids section for the unit. Along these lines, fire service publications often supply a listing of tapes, films, instructional booklets, plans, and informational guidelines helpful for providing the specialized training for the unit.

Disaster Drills

Multiagency, large-scale disaster drills, any of which require the special equipment and expertise of rescue firefighters, present rescue company members with an opportunity to participate actively in training sessions and to exchange ideas and suggestions on topics including various ways to use special tools and methods for evaluating a department's equipment needs. Many operational procedures used at large-scale incidents, in fact, have been developed through training at such drills.

The Firehouse

A firehouse can be used for training (Figure 7.7). Previous rescue incidents can be critiqued; and, if necessary, operational procedures can be corrected, expanded, refined, and practiced at quarters. Lowering and lifting "victims" and stretchers, for example, can be done from the build-

Figure 7.7. Firehouse roof being used in rope-training session. PHOTO BY AUTHOR

Figure 7.8. TOP: *This automobile "air bag" was removed from a vehicle used in a firehouse training drill.* PHOTO BY AUTHOR

Figure 7.9. BOTTOM LEFT: *As a young firefighter, the author acquires first-hand knowledge about being a "victim" during a drill.* COURTESY RESCUE COMPANY 2

Figure 7.10. *Going back to the site of an operation provides greater understanding when training firefighters who did not actually operate at the incident.* COURTESY S. SPAK

ing's roof or setbacks. Many units have collected various objects that were part of a rescue operation; brought them back to the quarters (Figure 7.8); and, through training with various tools and equipment, developed standard operational procedures for future use.

A picture hanging in my den still vividly brings to mind, even though it was more than 20 years ago, an in-house training session held during a night tour in my early days with the rescue company. My officer, a salty old lieutenant, decided that I would be a "victim" (Figure 7.9) and my fellow firefighters would secure me in an army litter. Lying on the litter, I tried to listen to the lieutenant's every word and direction concerning the proper lashing of the "victim" to the stretcher. My "brothers"—with straight faces—neatly and with great proficiency secured me to the stretcher so that I could be lifted horizontally or vertically. The pole hole was to be used for the vertical lift on a short ladder. Ropes for the lift were lowered, the guidelines were placed, and slowly I was lifted halfway up the pole hole. The lieutenant gave the signal, and the dinner bell rang. I watched in amazement as the brothers left to partake of the evening meal. The picture hanging on my den wall was taken by one of the brothers that night. Rest assured that in later years I really savored my meal while the new guy lay on the stretcher pondering his role as "victim." Incidentally, we used that lift in a number of operations, and it always worked well. Its success could be attributed not only to the imagination the lieutenant brought to the training sessions, but also to the scheming brothers' participation and interest in ensuring that the knots were properly tied. It worked every time.

The Rescue Site

Returning to the site of a rescue incident is another form of training (Figure 7.10). Critiquing at the site with safety in mind gives a better understanding of the operational procedure. I've listened to many a scenario of rescue operations, but I did not fully understand them until I personally had visited the sites.

Rescue units often are called on by the department to share in and to evaluate a proposed piece of equipment or a special procedure. Rescue company members' well-rounded experience and expertise are valuable assets for fine-tuning a proposed operation or for changing the specifications for experimental equipment. These activities are training sessions in disguise that add to the knowledge the unit takes to an operation.

Inspections and Operational Checks

Regularly scheduled checklist inspections and maintenance operational checks also can be training activities (Figure 7.11). These procedures should be structured as a lesson. Operational performance checks on the air-bag system, for example, include the following evaluations: the bag's placement (ensuring that the various sizes (10) and shapes (two) are in their correct positions), hoses, manifolds, regulators, adapters, and air-supply replenishment; these combined checks constitute a training session. Likewise, by involving all on-duty members in the observations, operations, and discussions of the equipment's uses, limitations, and capabilities, the equipment inspection becomes a drill session. This occasion also is an excellent time for experienced rescue unit members to become actively involved with the training and to pass on many of the tips and hints that have been "grandfathered" down to new members.

Training must include everyone, from the newest member to the most experienced members, officers included (Figure 7.12). Officers? Yes. To lead a specialized group of firefighters, the leader must know at least the capabilities and limitations of the tools and equipment and their proper placement.

That sounds good for the regularly assigned officer who is involved in the day-to-day activities and training sessions, but what about the fill-in or the officer assigned on a temporary basis? This individual must be given some familiarization training. The unit's training schedule must take these officers into account, and training sessions should be geared to the officers as well as the firefighters. While the firefighter is involved mostly in practical applications, the officer is concerned with the theory and principle of the equipment's operation.

Training for the new rescue company members has to be spread out over a long period of time, mainly because of the vast amount of equipment these specialized units carry. Again, the unit's needs and objectives determine the how, what, where, and when of the new members' training. Having a specialized training school in the department is the ideal; it makes it easier to build a basic training foundation. Realistically, however, few department budgets provide for this luxury.

In-House Programs

The aim of an in-house program should be to teach members the principles of operating the rescue company's specialized equipment and

Figure 7.11. *The training session includes a maintenance and operational check of the equipment used.* PHOTO BY AUTHOR

Figure 7.12. *Training and drill sessions should include officers and all members.* COURTESY J. REGAN

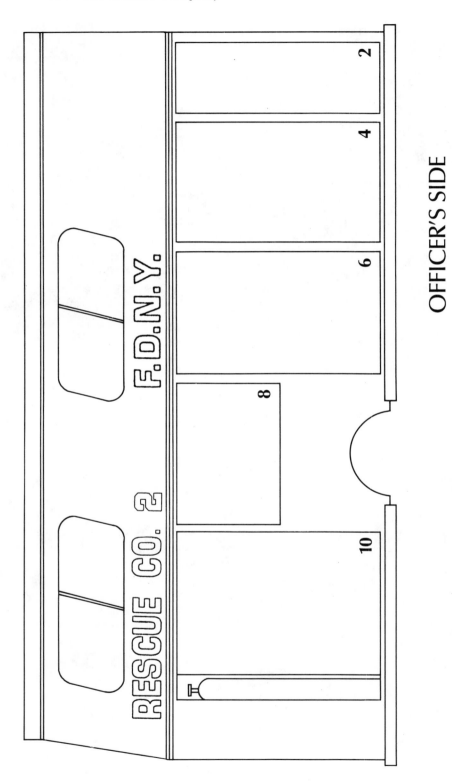

OFFICER'S SIDE

Figure 7.13. A blank schematic of the apparatus compartments can help train new members. New members should be required to list all the tools and equipment in the compartments.

tools and their practical uses and to help them become experienced in these uses. The first step in establishing this type of program is to develop a training format (Figure 7.13). Tool usage based on a company's response records can help accomplish this goal. If, for example, a company's records indicate that a hydraulic spreading device was used more often than any other piece of equipment over a six-month period, the training program for new members may begin with that device. Similarly, each tool and piece of equipment the company used should be listed and that list should be a guide for scheduling the training sessions. The training schedule should coincide with the members' work schedules.

Some departments schedule training sessions for a particular time during a tour, a practice that helps to standardize policy. In a special unit, such as a rescue company, training can be done at any hour. The availability of an abandoned, burned-out vehicle, for instance, could provide the components for a drill on spreading devices and cutters, whether it's morning or night; rescue operations occur at any hour. So can training sessions. Just as the rescue unit is prepared to respond to operations whenever and wherever they occur, so should it participate in some "nonscheduled" training sessions that arise because the opportunity (such as a vacant house or abandoned vehicle encountered on the way back from a response) presents itself.

We've looked at the why, where, and when of the rescue company. In the next section, we'll take a close-up look at the *how*.

Figure 7.14. *Firefighters from New York traveled to Birmingham, England, to participate in a rope-rescue exposition.* COURTESY CITY OF NEW YORK FIRE DEPARTMENT

PART TWO

Rescue Incidents

8

IMPALEMENT

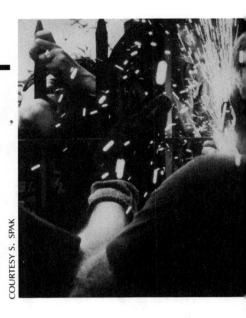

COURTESY S. SPAK

IN THE RESCUE FIELD, you often hear, "That sure was an odd job." The unexpected has become the expected. When we respond to serious vehicular or industrial accidents, we usually don't envision some of the "odd jobs" that seem to be happening more often. A number of years ago, my unit responded to a particular "odd job" five times in one year. All five involved impaled victims. This degree of frequency for this unusual type of incident has not been repeated since that time.

What I remember most about these incidents are the valuable lessons we learned and were able to use at any type of rescue operation later encountered. These out-of-the-ordinary operations also help to prove the value of training and critiquing and then adjusting and improving operations as necessary, based on the lessons learned.

The unusually high frequency of similar "odd jobs" over a short period of time impresses many of the lessons learned upon the "mental computer" (the one under the helmet) for a long time afterward. The first and the fifth incidents of the five impalements that occurred during that year ironically involved the same picket fence (Figure 8.1). Both operations were almost identical, with one exception: We had the advantage of incorporating all the lessons learned from the previous four incidents into the fifth incident.

Figure 8.1. *This picket fence was involved in two almost identical impalements.* PHOTO BY AUTHOR

Incident One

The first incident involved a young man in his twenties who either fell or jumped from the upper floors of a seven-story building and impaled himself on a one-inch-square, six-foot-high steel picket fence that surrounded the property. The victim had pickets in his back, leg, and arm, and he was still alive. The primary concern was to provide support so that the weight of the victim's body would not exacerbate his condition.

Rescuers first placed under the victim a full backboard supported by six-foot hooks fixed horizontally under the board and held by members. In this way, some of the pressure on the victim was relieved. Paramedics arrived early in the operation and immediately initiated victim-stabilization as rescuers were busy removing the victim.

Cooperation between medical personnel and rescuers during these incidents is a must (Figure 8.2). Understanding the duties and responsibilities of the other agencies is very helpful during these types of operations. Rescue company personnel placed a portable backpack cutting torch into operation. The victim was covered with fireproof blankets, and horizontal and vertical cuts were made to separate the section of the fence with the victim from the main body of the fence. A handline was stretched, charged, and standing by, should any sparks cause minor fires. Additional manpower had to be provided for support, since each cut shifted body weight from the fence to the rescuers. The cut sections helped to relieve pressure on the victim. The section cut was small enough so that weight would not increase too drastically, and far enough away from the victim so that he would not be endangered.

An ambulance was waiting. Its team made special accommodations for the victim's condition and for his very large size. Ambulances usually are not designed for oversized passengers (Figure 8.3), and the rescuers were strategically positioned inside the ambulance to assist in this most delicate transportation. Rescue company members preceded the ambulance to the hospital and briefed hospital personnel on what to expect—these incidents don't fall into the category of routine emergency room visits. With a closely coordinated team effort, the victim successfully was moved to the hospital.

Figure 8.2. Cooperation between medical personnel and rescuers is a must. COURTESY S. SPAK

Figure 8.3. *The design features of ambulances normally do not account for "oversized" passengers.* COURTESY S. SPAK

Incident Two

The second impalement incident that year involved a six-year-old boy attempting to climb the picket fence surrounding a high school. He lost his balance and impaled himself in the left temple. The incident occurred near the quarters of a fire company, and by the time the rescue company arrived, members already were supporting the victim, attempting to prevent further injury. Two members, held up by portable ladders and other firefighters, were able to straddle the fence and provide the necessary victim support.

Rescuers, using a cutting torch with great care, skillfully cut a section of fence from the main body of fencing. As the boy was removed from the fence and placed in the ambulance, the spike worked itself loose and came out of his head, an occurrence attributed to the boy's restfulness while awaiting help and the relaxation of all his muscles after he had been successfully removed from the fence.

Incident Three

The third incident involved a preteen girl who accidentally fell on a picket fence while roller skating. She landed on a one-half inch spike that penetrated her chin. Instinctively, she flexed her head back, thrusting the spike into her mouth (Figure 8.4).

Figure 8.4. *In a number of incidents, the victims flexed their head back and the spike exited through the mouth. Remarkably, no major damage occurred to the neck, chin, or throat.* COURTESY S. SPAK

The first-arriving unit members found her in this position and were awed by her composure. This calm was evident throughout the operation. The unit had requested the services of a rescue company and went about providing support, comfort, and the psychological reassurance needed during these types of operations.

After assessing the situation and providing additional reassurance, surgeon-like cuts were made using a portable cutting torch. The fence and young lady were carefully removed by ambulance to a special trauma team awaiting her arrival at the hospital. The position of the fence made it necessary to transport her face down and with padding between her body and the fence to preclude the possibility of her choking on her own blood. Again, a closely coordinated team effort was necessary.

Incident Four

Just two weeks later, a fourteen-year-old fell from her second-story window and was impaled on a picket fence used to prevent anyone from falling into the cellar entrance and stairs. Fire department units first at the scene found two family members supporting the victim on this four-foot-high fence. The lower height of the fence made supporting the victim much easier than during the previous impalement operations. Much of the cutting of the fence was done before the rescue company arrived. Only a few spikes remained to be cut before the victim and the cut section of fence could be separated from the main portion of the fence. Experience has shown that teamwork is the key at this point of the operation. The hospital was only a few blocks away, and members prepared to transport the victim in the same careful manner used in the previous incidents. The lessons learned and the previous experiences helped rescuers and hospital personnel.

Incident Five

Déjà vu. Just one year after that first impalement incident, someone fell off the same building and on to the same picket fence—a 19-year-old woman who plunged three floors and was impaled in the leg, buttocks, and abdominal area. As fate would have it, the rescue company was taking up from another incident and was just three blocks from the scene.

As in the four earlier impalement cases, additional manpower was needed to support the victim. She was suspended in air, face up. Since a larger section of the fence had to be cut, two cutting torches were used. Two members who had worked at the previous incident involving the same fence worked this tour; their experience was invaluable.

Because a large portion of the fence had to be cut, transporting the victim involved additional problems. After clearing the ambulance, the victim, fence, and much-needed manpower (to support the fence) just barely fit into the ambulance (Figure 8.5). Members went ahead to the hospital to make sure that the fence would clear the entrance doors and elevators. Making certain that the operation would not be interrupted by some unforeseen obstacle was part of our updated game plan, which contained the lessons learned from previous incidents.

Figure 8.5. *The fence is cut so that a section can be removed without causing further injury to the victim. The victim and the portion of the fence still within the body are transferred to the hospital where the fence is removed in the operating room.* COURTESY S. SPAK

Reviewing the Operation

Our company policy of reviewing and discussing our special operations report each tour is an important mechanism for passing the lessons learned on to all company members. The officer in charge of the incident fills out an in-house report, listing the equipment used and the particulars of the incident, giving special attention to the lessons learned. At the beginning of each tour, all members review these reports; and any member on duty who worked the incident adds his/her personal experience to the group discussion.

After the first incident, we took the section of fence involved in the incident from the hospital (with the approval of all parties involved). Cutting and testing various types of fences helped us to develop and upgrade our SOP for impalement operations. We devoted some of the training sessions held in quarters to having members use the fence to test and compare various power tools with regard to speed, ease, handling, noise levels, and overall effectiveness (Figure 8.6).

Figure 8.6. *A portion of a fence taken from one of the incidents is used in a training session at quarters to compare the cutting capabilities of various power tools.* PHOTO BY AUTHOR

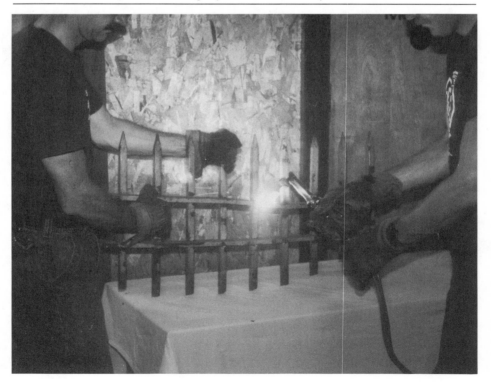

We used the cutting torch in all of the incidents for the following reasons:

- It provided better control and maneuverability than the power saw or sawsall.
- It cut faster than the power saw or sawsall.
- It eliminated and reduced the noise level in comparison with the power saw. Noise can be a very important psychological factor for a victim. The victims in all five incidents were conscious, were aware of what was going on around them, and could hear rescuers discussing various aspects of the rescue operation.
- Test results showed that heat conduction was not a problem, based on the speed of the cutting and density of the material.

Lessons Learned

The following list of lessons learned can help to make impalement rescues successful.

- When using torches and cutting equipment, all potential hazards must be removed from the immediate area.
- A protective fireproof blanket must be provided for the victim. Turnout coats can be substituted, but not the coats protecting the members doing the cutting.
- A charged handline must be ready, and rescuers supporting the victim must be made aware of all cutting operations and be alert to the possibility of sparks.
- The rescue operations officer (ROO) must reach into the "computer" (the one under the helmet), mentally print out the checklist, and cover all the bases.
- The fence or other object on which the victim is impaled should be removed by hospital personnel, who are better equipped than personnel in the field to deal with complications such as cardiac arrest or uncontrolled bleeding. Only at one of the above incidents did an inexperienced emergency worker suggest removing the fence at the scene.
- Members should precede the ambulance to the hospital to provide the medical staff with information regarding the patient's condition, the type of fencing involved, and the areas of penetration and impalement. They also should check for clearances that can accommodate oversized patients or large sections of fence or other

objects not yet removed from the victim and have the entrance area, hallways, and emergency rooms cleared of patients, visitors, and unneeded hospital personnel. Although emergency room personnel see numerous accident victims day in and day out, seeing an impaled victim can upset them, to say the least. In two incidents, volunteer workers openly expressed shock at seeing the accident victims, which was not overly reassuring to the victims.

- The hospital personnel in charge should clear all workers except those actually needed.
- Rescuers should be prepared to scrub down, if necessary. In these operations, rescuers were standing by in operating-room gowns in case special tools or equipment were needed to help remove the fence.
- All the lessons learned should be combined and used to upgrade the operational plan and develop SOPs so that operations like these will not seem as "odd" the next time.

The expertise of the rescuers and medical personnel at the scene and in the hospital were major factors in the survival of four out of the five victims of these traumatic incidents.

9

OPERATING ON AIR

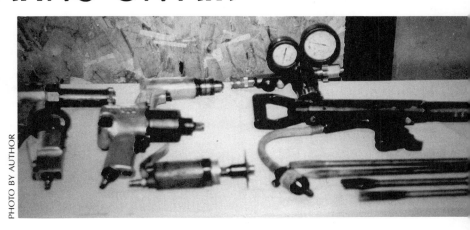

PHOTO BY AUTHOR

A NUMBER OF POWER SOURCES for operating tools and equipment during rescue incidents are available. Hydraulic, electric, fuel-powered, manual, and air are the most common. When the term *air* is discussed in the fire service, it usually is associated with self-contained breathing apparatus. For rescue companies, the term *air* has many more meanings and uses. Air chisels, air-impact hammers, air-operated circular saws and drills, demolition hammers, pavement breakers, rotary drills, impact wrenches and sockets, air bags, leak-sealing bags, airshores, air jacks, Wizzer saws (discussed in detail in this chapter), air extension lines for SCBA, air supply for scuba operations, and special fittings for supplying air through hose to reach trapped firefighters are just some of the possible ways air can be used to supply power for tools and equipment. The air can be supplied from SCBA cylinders, larger 300-cubic-foot cylinders, compressors built into the apparatus, portable compressors, or commercial-size compressors.

Air-Powered Tools

The New Zealand Fire Service, when faced with what management believed to be cost-prohibitive expansion plans, came up with its own solution for the problem. In "Working on Air" in the December 1988 issue of *Fire Engineering*, the author, Kevin O'Sullivan, describes the benefits derived from working with air-powered tools: "Air is clean, requires only a single feedline to the tool, is intrinsically safe, and is totally reliable in starting. Air-driven tools are simple, lightweight, and powerful. They are also relatively quiet" (Figure 9.1). *Quiet* was the word that put my personal "computer" (the one under the helmet) into action.

The Wizzer Saw

Many of these air-operated tools have been employed in the commercial, industrial, and automotive fields for years (Figure 9.2); the Wizzer saw, for example, is used in the automotive industry. The fire service adapted it for its unique needs, as it has other tools.

Weighing only two pounds, the Wizzer saw is easily handled by one firefighter. The tool can be used for two to three minutes operating at 90 psi from an SCBA cylinder with regulator. A three-inch carborundum cutting blade turns at 20,000 rpm; a lexan guard covers the blade area, providing some protection for the tool operator. The tool can be used on case-hardened locks, sheet metal, and other materials, depending on their thicknesses. Testing, training, and drilling with the tool provide the necessary information regarding that particular tool's capabilities and limitations. Various brands, offering different capabilities, are available.

The Wizzer Saw and the Doorknob Lock

A few years ago, my unit was called to assist in the removal of a four-year-old boy's thumb from the locking device of a doorknob for a bathroom door. Apparently, while trying to lock the door, the child pressed in the center button, which then collapsed into the door handle and assembly; his thumb became jammed in the tiny knob hole.

Family members' efforts to remove the thumb only aggravated the situation. After having used butter, lard, and ice to try to free the digit, a family member's tugging caused the thumb to swell more and become securely wedged into the tiny hole. During efforts to free the thumb, a retaining pin that normally releases the inner and outer doorknobs became dislodged, further complicating the situation. Every attempt to separate the knobs caused the boy more pain.

Truck company members at the scene cut the entire lock assembly

Figure 9.1. *Air tools can be supplied from a variety of air sources.* PARATECH, INC.

Figure 9.2. *Many of the air tools used by rescue units have been used in the commercial, industrial, and automotive fields.* PHOTO BY AUTHOR

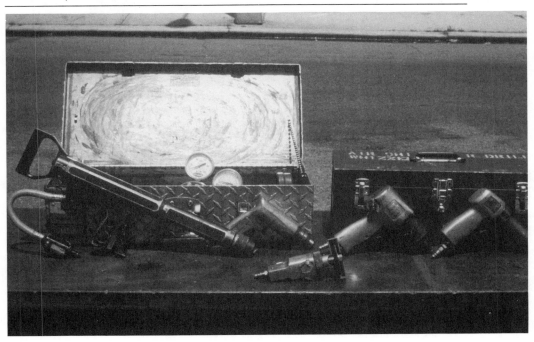

out of the door, hoping to make it easier to free the thumb. Promises of a ride on a big, red, shiny fire truck helped to calm the hysterical youngster, who thought the large power saw was being used to free his thumb instead of cutting away a section of the door.

My unit had been requested to bring our Wizzer saw to the scene. The incident commander, however, taking into consideration the effect that using another cutting saw would have on the child's emotional state, decided it would be better for professional medical personnel to assess the situation. Consequently, the hospital staff was notified.

At the hospital, the youth's hand was anesthetized, and a trauma specialist made several attempts to free the thumb. The doctor, when told of our Wizzer saw, agreed to let rescue members make cuts along the knob assembly in hopes of freeing the thumb. The tool was set up, sufficient air was available from the SCBA (Figure 9.3), and members awaited the doctor's instructions. Before starting the saw, all flammable items were removed, a safe environment was established, and hand extinguishers were readied for use. (Surprisingly, sparks did not become a problem.)

Figure 9.3. Wizzer saw supplied from SCBA cylinder. PHOTO BY AUTHOR

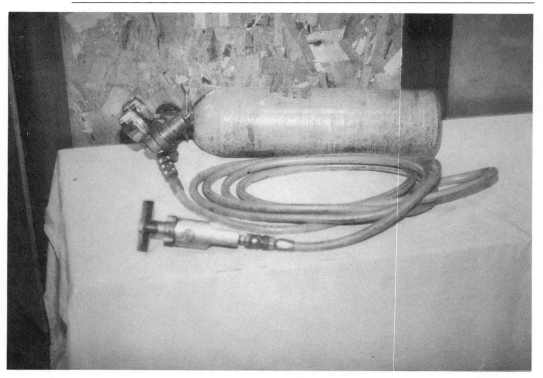

The boy became more apprehensive upon seeing the tool, regulator, hoses, and air cylinders. Psychological reassurance must always be one of the resources rescue members have in their arsenal: This was the time for heavy artillery. How do you gain the confidence of a hysterical four-year-old?

I thought back to the time when I took my son for his first haircut. I spent most of my time that memorable day acting like a middle line-backer, stopping my young boy from trying to escape from the barber shop. The barber's solution was to let him play with the comb and brush, listen to the scissors open and close, and make airplane-like noises, all the while cutting a lock of hair at a time. Why not let this youngster feel the tool, listen to it, and watch one of the firemen with the tool—similar to what the barber had done with his scissors?

Much to the surprise of the boy and the medical personnel, the tool's sound was much less noisy than had been anticipated; it sounded like a state-of-the-art dentist drill.

After a few minutes of "show and tell," it was time for the real thing. The youth's hand was held firmly as precise cuts were made. After the cover of the knob had been removed, the thumb was exposed. It had been bent down into a narrow opening as the youngster had applied pressure to the lock assembly. With a little manipulation, the doctor was able to free the thumb.

The incident helped us develop a standard operating procedure (SOP) for our Wizzer-saw operations. Company drills helped acquaint company members with the tool's capabilities and limitations, working pressures, and the amounts of air consumed by the cutting of various types of materials. Discussion of special operations involving air-operated tools, including the Wizzer saw, also were part of the drills. The lessons learned from these sessions and the training drills involving these tools made it possible to resolve successfully two almost identical incidents.

The Wizzer Saw and the Meat Grinder

At first mention of a meat grinder, my "computer" (the one under the helmet) spits out a "graphic printout" of my grandmother's hand-operated meat grinder from the early 1900s. While responding to a report of a person with his hand stuck in a meat grinder, my first thoughts were, Has anybody tried to disassemble the machine? Couldn't the hand be worked out by using butter or lard? But, the next two incidents involved commercial meat grinders that resembled my grandmother's grinder only in name.

INCIDENT ONE

The first incident involved a 14-year-old who had been working with his father in the meat section of a supermarket. When rescuers arrived, they found the youth's hand and part of his arm enmeshed in the meat grinder. Medical personnel on the scene monitored the youth during the entire rescue effort. A two-member team—a tool operator and another member acting as the "eyes and ears"—set up the Wizzer saw, prepared it for use, discussed the operational plan, and provided the necessary psychological reassurance.

First, horizontal cuts were made along the barrel section and then vertical cuts intersecting the horizontal cuts were added (Figure 9.4). An additional cut provided a window opening into which the tips of a

Figure 9.4. *This series of cuts made with the Wizzer saw and a porto-power spreading device helped free the hand of a 14-year-old boy from a supermarket meat grinder.*

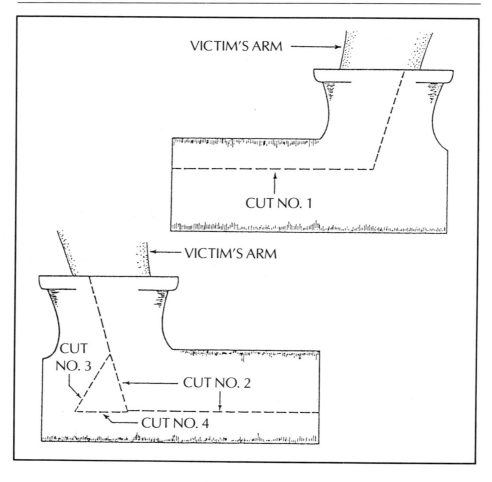

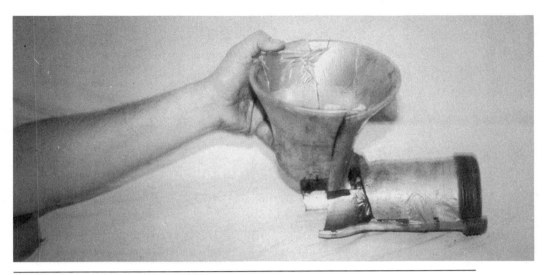

Figure 9.5. This meat grinder, identical to the one in the incident involving the 14-year-old boy, was cut horizontally and vertically with two Wizzer saws—one on each side of the grinder—to free the hand and arm of a middle-aged man. This "two-tool operation plan" eliminated the need for additional cuts and the use of porto-power. PHOTO BY AUTHOR

porto-power spreading device were inserted. The spreader was activated slowly and the cut sections separated. The worm gear had to be removed to free the hand completely. Because of the seriousness of the injury, the boy was transported by helicopter to a major trauma facility.

INCIDENT TWO

The second incident involved an identical meat grinder, but the victim this time was a middle-aged restaurant worker. The man said that while he was cleaning the grinder, a cleaning rag was sucked into it, dragging his hand and part of his arm with it. From the strong odor of alcohol on his breath, it was clear that the man had been drinking; a language barrier caused additional problems for rescuers. A bilingual firefighter was on hand to communicate with the victim in his native tongue, but the other problem couldn't be resolved so easily.

As instructions were relayed to the victim, firefighters made him as comfortable as possible while positioning the tool operators. Two Wizzer saws were at the scene, and a two-tool operation plan was formulated. By cutting on both sides of the grinder, horizontally and vertically, the need for the additional cuts and porto-power was eliminated. The intersecting cuts on two sides were all that were required to separate the grinder and free the victim's hand and arm (Figure 9.5). As in the first incident, the worm gear had to be removed to free the victim completely. Keeping the highly emotional victim calm was almost as difficult as performing the surgically precise cuts on the grinder. The talents of these specially trained rescue firefighters were never more evident than during this operation.

Planning Ahead

The printout (yes, the mental one) helped to ensure that sufficient air would be at the incident scene. We had learned that air was used for approximately 45 minutes during the first meat-grinder incident. We anticipated an extended cutting operation, so additional 300-cubic-foot air cylinders were requested; using two Wizzer saws would require additional air (Figure 9.6). Our apparatus was equipped with two 300-cubic-foot cylinders that could be used from the apparatus or brought by hand to the scene. As had been the case with the young boy in the hospital, the location of the second meat-grinder incident precluded using air directly from the apparatus.

Figure 9.6. A sufficient air supply must be at the scene. In some instances, such as the two meat-grinder incidents described previously, the location of the accident makes it impossible to obtain air directly from the apparatus. Carrying 300-cubic-foot air cylinders on our apparatus ensured that air was available for these extended cutting operations. PHOTO BY AUTHOR

Contingency plans are a vital part of a rescuer's game plan, as are the "what ifs": What if the hoses don't reach? What if we need more air? As the officer in command, I had to review the type of plan that would be implemented and to have a contingency plan on standby. Our associated engine company, which was familiar with our tools and the equipment stored in quarters, brought the additional air cylinders. Also, the incident commander had anticipated that personnel would be needed to bring the cylinders, hoses, regulators, and equipment as close to the scene as possible. All of these factors contributed to the success of the rescue operation.

Metal Chips a Hazard

An interesting sidelight to the operation involved the chips of cast steel being thrown off from the grinder during the sawing operation. While in quarters critiquing the incident, a member noted that the tool operator's work shirt had an unusual colored pattern outlined on it. On closer investigation, it was determined that the pattern was formed from the metal chips that had been thrown off while cutting the meat grinder. A medical examination showed that these chips had penetrated the member's skin. Because of the limited working area and the high temperature, most members removed their turnout coats. All members who had worked in the immediate area of the incident were required to undergo chest X-rays, which revealed that the chips had penetrated the chests of five rescuers. They were assured that most of the chips would be gone after a good shower and that they were in no immediate danger.

Lessons Learned

The following lessons were learned from these incidents:

- Proacting, rather than reacting, is needed. Sufficient air must be at the scene. Having too much air should not be a concern: It's cheap. Personnel always must be considered: The trauma room or restaurant work area may be on the upper or lower levels and unable to receive air directly from the apparatus.
- Maintaining tools and equipment at regular intervals helps ensure that they will be reliable during an emergency.
- The contingency plan, in addition to providing for having suf-

Figure 9.7. TOP:*This power saw was used many years ago to free a hand trapped in a meat grinder.*
Figure 9.8. *The state-of-the-art Wizzer saw used for the meat-grinder incidents described in this chapter is much smaller, lighter, and easier to control and operate than the power saw in Figure 9.7.* PHOTOS BY AUTHOR

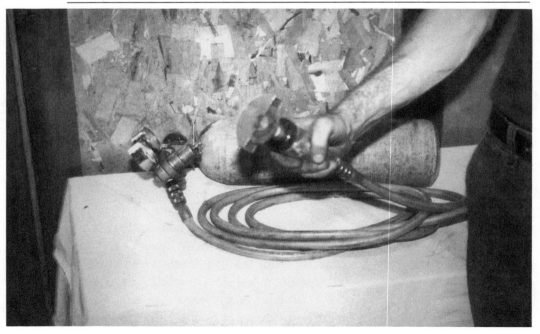

ficient air, personnel, and backup tools and equipment, also must take into account the need for improvisation.

● Donning proper protective clothing, including eye shields or goggles, is absolutely necessary.

Twenty Years Ago

A meat grinder incident that occurred 20 years ago was handled differently than the two incidents previously discussed—in accordance with the equipment available and in use at that time. A teenager's hand and arm were enmeshed in a commercial meat grinder. Hospital personnel requested the assistance of the rescue company, primarily for the power saw that had recently been introduced to the fire service (Figure 9.7). A power saw with a metal cutting blade was used as personnel stood by. The saw was held by two members and directed by a third member, the guide member. Two other firefighters held an extinguisher to cool the tool and grinder and to put out spark-related fires. The fumes produced by the saw necessitated a fan for ventilation. The high noise level of the saw created additional communications problems between rescuers and added to the victim's fears. Rescuers overcame these obstacles and successfully freed the victim's hand.

By comparison, we can see how the Wizzer's weight makes it much easier to operate (Figure 9.8). Only one operator is needed to hold and direct it. The air-operated Wizzer saw doesn't create the fumes that power saws do. Finally, and this is a very important factor, the much higher noise level generated by a power saw induces fear, thereby increasing stress for the victim.

As we have seen, whether it's the 1960s, 1970s, 1980s, or 1990s, the people or problems don't change—but the tools and equipment most certainly do. Using our specially trained and talented rescue personnel and their ability to improvise and adapt always will be one of the greatest strengths and assets of the fire service.

10

BURIED VICTIMS

COURTESY H. EISNER

THE ACTIONS RESCUERS TAKE at the scene of a buried-victim operation often determine whether the operation will result in a rescue or recovery. Rescue units often are called to the scene of an accident in which victims are buried (either completely or partially) as the result of cave-ins, trench or excavation collapses, and landslides. These accidents can be caused by a spoil pile (the excavated material) sliding back into a trench or opening; the ground beneath the pile giving way; vibrations from nearby equipment, railways, subways, trains, or traffic; improper sloping or undercutting of the walls; and water seepage that saturates the soil and causes it to become unstable. The leading causes of most cave-ins and collapses involving buried victims are the failure to follow safety rules and to provide adequate shoring and sheeting (or failure to provide shoring and sheeting at all).

According to the requirements of Occupational Safety and Health Administration (OSHA) section 1926.652— "Protection of employees in excavations: (1) Each employee in an excavation shall be protected from cave-ins by an adequate protection system designed in accordance with paragraph [b] 'Design of sloping and benching' (Figure 10.1) or [c] 'Design of support systems, shield system and other protective systems' (Figure 10.1). The exceptions to these rules are: *(1)* when excavations are made entirely in stable rock or *(2)* when excavations are less than 5 feet (1.52 m) in depth and examination of the ground by a competent person provides no indication of a potential cave-in."

Table B-1
Maximum Allowable Slopes

Soil or Rock Type	Maximum Allowable Slopes(H:V)[1] for Excavations Less Than 20 Feet Deep[3]
STABLE ROCK	VERTICAL (90°)
TYPE A[2]	¾:1 (53°)
TYPE B	1:1 (45°)
TYPE C	1½:1 (34°)

NOTES:
1. Numbers shown in parentheses next to maximum allowable slopes are angles expressed in degrees from the horizontal. Angles have been rounded off.
2. A short-term maximum allowable slope of ½H:1V (63°) is allowed in excavation in Type A soil that are 12 feet (3.67 m) or less in depth. Short-term maximum allowable slopes for excavations greater than 12 feet (3.67 m) in depth shall be ¾H:1V (53°).
3. Sloping or benching for excavations greater than 20 feet deep shall be designed by a registered professional engineer.

Figure B-1
Slope Configurations

(All slopes stated below are in the horizontal to vertical ratio)

B-1.1 Excavations made in Type A soil.

1. All simple slope excavations 20 feet or less in depth shall have a maximum allowable slope of ¾:1.

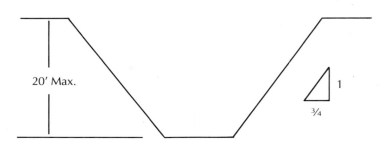

SIMPLE SLOPE—GENERAL

Exception: Simple slope excavations which are open 24 hours or less (short term) and which are 12 feet or less in depth shall have a maximum allowable slope of ½:1.

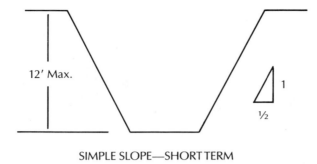

SIMPLE SLOPE—SHORT TERM

2. All benched excavations 20 feet or less in depth shall have a maximum allowable slope of ¾ to 1 and maximum bench dimensions as follows.

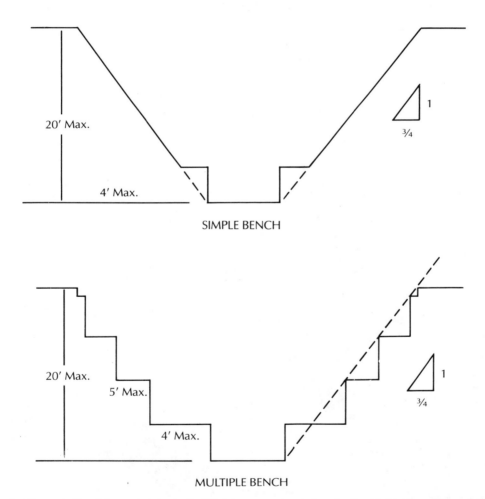

SIMPLE BENCH

MULTIPLE BENCH

3. All excavations 8 feet or less in depth which have unsupported vertically sided lower portions shall have a maximum vertical side of 3½ feet.

Figure 10.1. COURTESY OCCUPATIONAL SAFETY AND HEALTH ADMINISTRATION (OSHA)

Information Needed

When responding to the scene of a buried victim incident, rescue officers must obtain as much information as possible. This knowledge helps them to develop a mental picture and set the "computer" (the one under the helmet) in operation. On arrival at a construction incident, they must find the person in charge and get as much information as possible concerning the circumstances surrounding the incident (Figure 10.2). These facts include the following:

- How many victims are buried or unaccounted for?
- How long have they been buried?
- What are their exact locations? or Where were they last seen working?
- Are there any eyewitnesses—fellow workers, supervisors, equipment operator, or the like?
- Has there been any communication with the victim (for example, did a worker trapped in a large pipe indicate his location by tapping out sounds)?

Figure 10.2. *It is important to gather information relating to the incident from workers and witnesses.* COURTESY S. SPAK

Figure 10.3. Trenches requiring dewatering operations can greatly influence ground conditions. PHOTO BY AUTHOR

- What type of work was being performed (sewer, gas, or electric, for example)?
- What are the depth and width of the excavation?
- Are any plans or blueprints available?
- What actions have been taken so far?
- Are there any hazards rescuers should know about (such as high voltage or pipelines)?

Utilities a Problem

OSHA 1926.651 dictates that, prior to opening an excavation, a determination must be made as to whether underground installations (sewer, telephone, water, fuel, or electric lines) will be encountered, and, if so, where they are located. When the excavation approaches an installation's estimated location, its exact location must be determined; and after uncovering it, it must be properly supported. Utility companies must be contacted and advised of the proposed work prior to the start of the actual excavation.

Hazards such as electric, gas, water, pipelines, and sewers can add to an already dangerous condition. Trenches that require de-watering operations can greatly influence ground conditions (Figure 10.3). The danger that sewer lines pose should not be underestimated. Sewage treatment can produce methane gas, often found in sewer systems. Methane gas is potentially explosive; and because it's also odorless and colorless, it is extremely hazardous. Rescuers often must rely on contingency plans to overcome these obstacles.

The Operation

PERSONNEL

Supervisors, foremen, equipment operators, and workers must remain at the site to provide additional information that rescuers may require as the operation progresses. These workers cannot be depended on to help in the rescue, since many of them may be in a state of shock and unable to operate equipment or tools effectively. When the need for special equipment is anticipated, additional qualified operators should be at the scene. The number of rescuers needed varies according to factors such as the number of buried victims, the type and depth of the excavation or cave-in, the hazards encountered, and, most importantly, the type of emergency equipment that responded to the scene. In any emergency situation, additional help should be called without hesitation if it appears it might be needed. It is easier to return unnecessary personnel and equipment than to have to wait for help in a worsening emergency situation. "Proact rather than react" is a saying we often hear. Always having a backup rescue team standing by at a collapse operation when the possibility exists that secondary cave-ins or collapses might trap the primary rescue team is proacting. The backup rescue team should not be involved in the initial operations and should stay out of the danger zone.

The work area and rescuers must be made as safe as possible, and a minimum of rescue workers should be used in the danger zone. Ladders should be placed at both sides or ends of the excavation for entry and exit where feasible.

Rescuers should have safety lines attached to them during operations, and they should be rotated frequently to prevent fatigue and possible injuries. Working at these incidents is physically and psychologically demanding. This fact must be kept in mind.

The remaining walls must be supported before rescuers enter the trench or opening.

SHORING

Walls of shoring and sheeting material should be constructed down to the level of dirt during the digging phase. As dirt and material are removed, additional shoring should be continued. As rescuers remove two additional feet of dirt or material, for example, they should support the exposed two feet of wall with additional sheeting and shoring. Any lines or pipes exposed during the digging also must be supported (Figure 10.4).

In situations involving underground piping or lines, rescuers must be aware that the ground has been excavated before and that the walls could be unstable. A ladder placed across the opening will secure ropes supporting the piping or lines. The ladder must be placed on ground pads—sheets of ⅝- or ¾-inch plywood placed alongside the opening that help to distribute the weight, thereby preventing a secondary collapse.

SAFETY MEASURES

The ability of a partially buried victim to communicate should not deter rescuers from using proper safety measures or from shoring and sheeting the opening for support (Figure 10.5). Rescuers at times bypass safety procedures in their haste to free a victim. Having discipline, taking the proper safety measures, realizing the dangers to themselves and to the victim, and operating with caution help rescuers to prevent secondary collapse or cave-ins and prevents them from becoming victims themselves.

DIGGING

It is best to use the hands or small shovels when digging for victims. The trenching tool, with which military veterans are quite familiar, is ideal for these incidents. Because of its small size, it provides the versatility needed in small spaces where shoring often interferes with operations. The rescuer must be careful not to cause additional injuries to the victims when digging with a shovel. The shovel should be used to dig as close to the victim as safety permits. The remainder of the digging should be done by hand. Rescuers should use small pails or buckets, lowered by rope into the hole, to remove excess dirt or soil to a safe area. The removed dirt or soil should be placed far enough away from the opening so that it does not cause a spoil-pile slide, which can occur

Figure 10.4. *Lines or pipes exposed during the digging must be supported.* PHOTO BY AUTHOR

Figure 10.5. *Proper shoring is a safety measure that must not be bypassed.* PHOTO BY AUTHOR

when the soil, dirt, or material removed in digging is placed too close to the opening at a steep angle and slides back into the opening.

REMOVING THE VICTIM

When a buried victim is reached, a safety line (rope or harness) is tied to him or her. The victim must be completely dug out and never should be forced out by pulling with hands, rope, or machinery. The material surrounding the victim is compressed against the body, and pulling will only cause injury.

When uncovering a victim, the entire chest and head area must be freed to allow for chest expansion and respiratory exchange when providing oxygen. When the victim is completely free and ready to be removed, proper emergency care should be administered. The nature of these incidents cause many victims to suffer traumatic injuries. Again, haste must not cloud the priorities: The victim must be given the medical treatment needed before being removed. The "computer" printout checklist must include having medical personnel standing by during the rescue operation.

Often, because of the shoring or size of the opening, the victim must be taken out in a vertical position with a stokes or basket stretcher.

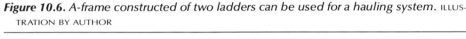

Figure 10.6. *A-frame constructed of two ladders can be used for a hauling system.* ILLUS-TRATION BY AUTHOR

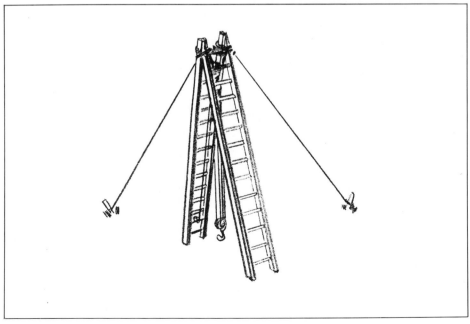

Securing the victim properly in the stretcher and closely coordinating the removal protects the victim against additional injuries. The type of lifting or hauling device used depends on the rescue company's equipment inventory. All rescue companies should carry these types of systems, but if none is available, pulley systems or a block and tackle can be rigged to portable ladders placed over the opening to provide a hauling or lifting system (Figure 10.6). An operation involving buried victims is not complete until the victim and the rescuer have been safely removed.

Making Excavations Safe

What are the requirements for making excavations safe? Government regulations require minimum dimension and maximum spacing standards for trench shoring (Figure 10.7). Some states have more stringent regulations that go beyond the minimum safety standards. Some contractors, however, still do not comply with the regulations, posing problems for rescuers in buried-victim removal operations.

My unit responded to a report of a worker buried in a sidewalk cave-in. What had really happened was that the contractor, working in a poorer section of the city, opened an excavation to repair a broken water line from the street to a multifamily dwelling. Although shoring material was only 25 feet from the opening, the contractor used no shoring or sheeting in the opening. One worker was killed when the side walls collapsed on him (Figure 10.8). Rescuers dug him out. Notification of the collapse was delayed (probably for from five to 15 minutes) as the contractor's crew attempted to rescue the worker before calling in the emergency. When contractors do not comply with the proper mandated safety rules, they sometimes attempt to rectify the situation before calling for help.

As the requirements for trench shoring indicate, safety standards are designed to protect the workers; and contractors who fail to comply with these requirements endanger not only their workers, but also rescuers who may be called to the scene.

Contractors' shoring of excavation sites usually involves using two- or three-inch-thick planks butted tightly together as uprights to form a solid sheeting, which is then held in place by wales (braces placed horizontally against the sheeting) and shoring. In addition, wales transmit loading from the sheeting to shores. Some major contractors also use solid, pre-formed panels, which are available in varying strengths to comply with government regulations (Figure 10.9). Because of the panels' sizes and weights, heavy equipment usually is needed to place

Timber Trench Shoring—Minimum Timber Requirements*

SOIL TYPE A P_a = 25 × 11 + 72 psf (2 ft Surcharge)

Depth of Trench (Feet)	Cross Braces							Wales		Uprights				
	Horiz. Spacing (Feet)	Width of Trench (Feet)					Vert. Spacing (Feet)	Size (In.)	Vert. Spacing (Feet)	Maximum Allowable Horizontal Spacing (Feet)				
		Up To 4	Up To 6	Up To 9	Up To 12	Up To 15				Close	4	5	6	8
5 To 10	Up To 6	4 × 4	4 × 4	4 × 6	6 × 6	6 × 6	4	Not Req'd	—				2 × 6	
	Up To 8	4 × 4	4 × 4	4 × 6	6 × 6	6 × 6	4	Not Req'd	—					2 × 8
	Up To 10	4 × 6	4 × 6	4 × 6	6 × 6	6 × 6	4	8 × 8	4			2 × 6		
	Up To 12	4 × 6	4 × 6	6 × 6	6 × 6	6 × 6	4	8 × 8	4				2 × 6	
10 To 15	Up To 6	4 × 4	4 × 4	4 × 6	6 × 6	6 × 6	4	Not Req'd	—				3 × 8	
	Up To 8	4 × 6	4 × 6	4 × 6	6 × 6	6 × 6	4	8 × 8	4		2 × 6			
	Up To 10	6 × 6	6 × 5	6 × 6	6 × 6	6 × 8	4	8 × 10	4			2 × 6		
	Up To 12	6 × 6	6 × 6	6 × 6	6 × 8	6 × 8	4	10 × 10	4				3 × 8	
15 To 20	Up To 6	6 × 6	6 × 6	6 × 6	6 × 8	6 × 8	4	6 × 8	4	3 × 6				
	Up To 8	6 × 6	6 × 6	6 × 6	6 × 8	6 × 8	4	8 × 8	4	3 × 6				
	Up To 10	8 × 8	8 × 8	8 × 8	8 × 8	8 × 10	4	8 × 10	4	3 × 6				
	Up To 12	8 × 8	8 × 8	8 × 8	8 × 8	8 × 10	4	10 × 10	4	3 × 6				
Over 20	See Note 1													

*Mixed oak or equivalent with a bending strength not less than 850 psi.
**Manufactured members of equivalent strength may be substituted for wood.

Figure 10.7. Occupational Safety and Health Administration (OSHA) requirements for soil type A-timber trench shoring. COURTESY OSHA

Figure 10.8. *One worker was killed when the unshored walls of this trench collapsed and buried him.* PHOTO BY AUTHOR

Figure 10.9. *Steel preformed panels used by contractors.* PHOTO BY AUTHOR

and remove them. When called to a scene involving a buried victim where the contractor has complied with the safety regulations, it is possible to determine the size, depth, and width of the trench or excavation by conferring with the person in charge or by using the blueprints or plans for the job. Often these job sites have additional shoring materials readily available should rescuers require them.

Additional Resources

More often than not, however, rescuers must improvise and make do with what they have at their disposal. The alternatives for rescue units when the necessary materials are not immediately available are the following:

- Compiling a list of suppliers (lumber yards, contractor-and building-supply yards, and so on) within or close to the response area and arranging in advance to obtain the materials needed for an incident by picking them up or having them delivered to the scene. Many contractors and suppliers are more than generous when asked to aid fire service personnel involved in such incidents.
- Including in the preplans a list of building and construction site locations in the response area. Many of these sites have materials that can be used for shoring or sheeting.
- Using the various tools and equipment carried by many heavy rescue units. Screw or trench jacks, jimmi jaks, and hydraulic and pneumatic shores usually are part of the rescue unit's inventory and can be used as trench shoring (Figure 10.10). Knowing the

Figure 10.10. Trench or screw jack, jimmi-jak, and pneumatic shores carried by rescue companies. PHOTO BY AUTHOR

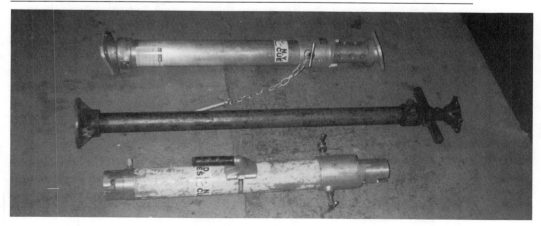

limitations and capabilities of these devices, however, is a must. It is recommended, for example, that screw or trench jacks not be used in trenches wider than five feet unless two of the jacks are used side by side. It is also good to know that air shores, which are activated by either carbon dioxide or compressed air, come in seven sizes, adapt to all types of trenches, and are approved by OSHA (Figure 10.11).

Alternate Resources

If the unit does not have the room to carry these tools and equipment, they should be stored in quarters or at a central location so that they can be delivered to the incident scene by another unit or by vehicles capable of carrying the equipment. A number of departments have formed specialized cave-in units, trench rescue units, and collapse units by using pickup trucks or other specially designed vehicles for this purpose. A

Figure 10.11. Pneumatic shore being placed in excavation. PHOTO BY AUTHOR

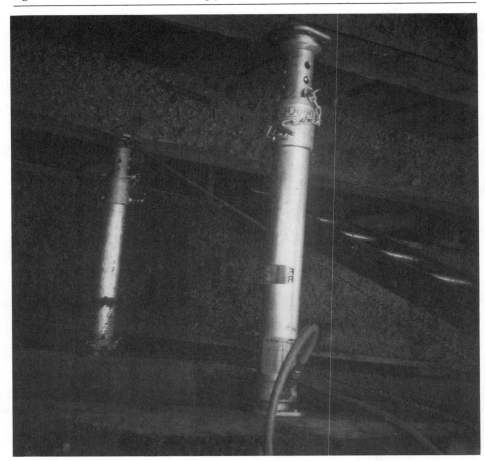

Figure 10.12. The inside of a specially designed apparatus that carries a variety of tools and equipment needed for incidents involving buried victims. PHOTO BY AUTHOR

very common method is to use platforms on demand (P.O.D.s). These containers, usually 16 to 20 feet long, are transported to an incident scene by truck. They can be used as command units, medical units, communications units, air systems, haz-mat units, cave-in or collapse units, or general-storage units. A department can make them multifunctional or store individual containers in a central location so that they easily can be brought to the scene by truck. In this way, the special apparatus or P.O.D.s can be brought to the scene as needed. Using P.O.D.s is a convenient and cost-effective means to provide a variety of special services.

Tools/Equipment

What type of special equipment/tools and materials should these units carry? Although the carrying capabilities of units vary, they must use their department's resources effectively (Figure 10.12).

Following is a list of equipment most commonly carried by rescue units.

EQUIPMENT LIST

- Plywood panels (4-feet by 8-feet, 1¼ inch thick), for sheeting

- Pre-formed panels, usually made of multilayered fiberglass (4-feet by 8 feet, 1 inch or ¾ inch thick), for sheeting
- Planking (2-inch by 12-inch by 12 feet long), for uprights
- Ground pads (4 feet × 8 feet × ¾ inch plywood)
- Timbers of various lengths (4 inches × 4 inches, 6 inches × 6 inches, 8 inches × 8 inches), for shoring
- Screw or trench jacks
- Air and hydraulic shores
- Jimmi jaks
- Saws: chain saws, power saws (wood, metal, and masonry blades), sawsall, circular saws, hand saws, hacksaws, heavy-duty drills and bits, and hole saws
- Hand tools: hammers, nails, sledgehammers, pinch bars, shovels, and trenching tools
- Plastic pails or buckets
- Rope of various sizes and lengths
- Slings and harnesses
- Generators, lights, extension cords, hand lights, electrical junction boxes
- Lineman's gloves and clampstick
- Dewatering pumps and hoses
- Utility blower and extension tube
- Monitoring equipment: oxygen, carbon monoxide, explosive meters, multifunctional meters
- Flares, traffic cones, and flags
- Ladders: straight and extension
- First-aid equipment: resuscitators, stretchers, and backboards
- Safety equipment: helmets, hard hats, gloves, lights, and safety glasses
- Tripods, hauling systems, pulleys, and block and tackle
- Cooler for water or energy drinks

This list is not all-inclusive; established units already may have many of these items and more. New units should stock up as best as their budget allows, but they should consult with established units for help in getting started. Ingenuity often makes up for a lack of material and equipment.

An important point to remember is that most buried-victim incidents are time-consuming operations. The game plan must use all available resources, ensure the safety of all rescuers, and include a contingency plan.

11

ELEVATORS

EVERYONE IN THE FIRE SERVICE has a story to tell about elevators. Experiences involving rescuing passengers from stuck elevators range from short, funny anecdotes to harrowing tales of dangerous operations during serious fires. Fire department operations involve both the emergency incidents and fire- or smoke-related operations complicated by the unit's having to use an elevator to reach the source of the fire or smoke.

Elevators create odd, frustrating, complicated, dangerous, and unusual problems. I vividly remember entering the hallway of an apartment building to find two residents holding on to the maintenance man who was dangling in midair with his arm stuck through the vision panel of the elevator hoistway door. The elevator had stalled, and he had been going to trip the release device to open the hoistway door and remove the passengers. As he reached in and moved his arm up toward the release device, however, the elevator suddenly moved upward; the man's arm moved with it, lifting him off the ground; and the elevator just as quickly stalled again.

After providing a makeshift platform on which he could stand, we worked to free him completely. Using a small air bag and working from above, we were able to move the elevator car and free the arm. We were surprised that he suffered only minor scrapes and cuts.

147

Deadly Game

Some teenagers, on the other hand, weren't so lucky. Since 1984, at least 11 have died and 70 have been reported injured (authorities believe many more injuries go unreported)—most of them from housing projects featuring high-rise buildings—as a result of playing a game they call "elevator surfing," "riding the el," or "being on top" (Figure 11.1). The game involves jumping back and forth between elevators, scaling the cables, and hanging from the girders in the shaft. The youngsters use a string or other hand-fashioned devices to release the locking bar on the elevator doors, usually springing the doors on the second floor while the elevator is stopped on the first floor. Then, they climb on top. They become victims by falling from cars, being crushed, or being struck, while in the shaft, by parts of the elevator machinery.

As a young firefighter, my unit frequently was called to housing projects for elevator emergencies. A seasoned veteran told me that the fastest way to get the hoistway door open in such a situation is to ask one of the local kids to help. In their idle time, these youngsters had perfected the technique of opening the doors so that they could go "elevator surfing."

Figure 11.1. *Warning signs on elevators aim at stopping youths from playing elevator "surfing," which has caused 11 deaths and 70 reported injuries in New York City during the past seven years.* PHOTO BY AUTHOR

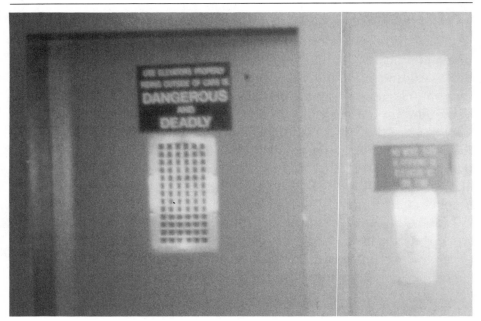

Recently, steps have had to be taken to prevent this easy entry. Many of the emergency key openings have been sealed over, and plates are used to cover the vision panel and other openings. Although such precautions have helped to curb illegal entry, they pose additional problems for firefighters during elevator incidents.

Two incidents—one occurring as recently as October 1989 and the other many years ago—demonstrate the need for all units to be trained for elevator incidents. During the San Francisco earthquake in October 1989, problems included collapsed buildings, fires (often fed by broken gas mains), gas leaks, water-main breaks, and numerous stalled elevators resulting from the major loss of power. Preplanning, recalling off-duty members, and using resources effectively were instrumental in enabling rescuers to deal with this highly visible and publicized tragedy.

Northeast Power Failure

In November 1965, a major power failure in the Northeast created some serious problems for firefighters. The Otis Elevator Company reported that it serviced approximately 20,000 elevators in the affected area at that time. A total of 355 passengers were stuck in 161 elevators in 107 buildings. Remarkably, most passengers were removed in less than one hour, except those stuck in express hoistways; their rescues took hours, depending on the availability of the rescuers. Of the 134 cars stuck in hoistways, 112 stopped close enough to the floor so passengers could be removed by opening the car and hoistway doors.

Among the obstacles encountered during these incidents were the initial lack of personnel to search for trapped occupants in thousands of elevators in numerous high-rise structures while hampered by a lack of lighting and the need to use the stairs to reach the elevators on the upper floors.

Elevator Types

The types of elevators encountered range from the simple single-shaft passenger or freight elevator to the glass, fully enclosed passenger elevators that run along the outside of buildings or in atriums. The "double-decker" cars now in use save space because they have the equivalent of two cars in a single shaft. The majority of elevators are single-car or multicar (Figure 11.2), and they provide local service (stop at each floor in a particular shaftway).

Figure 11.2. *Multicar elevators.* PHOTO BY AUTHOR

SINGLE-CAR AND MULTICAR ELEVATORS

They provide express service and usually ride the blind shaft for a number of floors before stopping at a floor. A bank of express elevators servicing floors 17 to 35, for example, rides in a blind shaft (no floor openings for service) from the lobby up to the seventeenth floor (the first stop). Elevators in blind shafts can present difficult challenges for firefighters.

SOPs Needed

Whether responding to an elevator incident in a high-rise office building or a housing complex, rescue units must have a standard operating procedure. As much information as possible must be acquired before and on arrival at the scene. Before an incident occurs, members can become familiar with potential incident sites through inspections. A building maintenance worker usually is available to provide members with the information needed to institute operations during an emergency such as the locations of the elevator machinery room and the elevator's main electrical power switch. Any special instructions regarding the building elevators noted during the inspection should be included in the company's records. If the department uses a computer-assisted dispatch system, the information should be programmed into the system.

If unit members cannot locate a maintenance worker when they arrive at the scene, they should try to find the person who called for assistance to determine the location of the elevator car involved in the emergency and the reason for the call. If the individual cannot be found immediately, rescue unit members must locate the elevator car by other means.

Elevator Emergency Procedures

LOCATING THE CAR

When entering the elevator lobby, rescuers should check the indicator panel to determine the location of the car involved. If the lobby has a telephone or intercom system with a capability to reach each elevator car, rescuers should try to contact the passengers and ask for their location.

Some hoistway doors have vision panels. Through these panels, rescuers can see if any part of the elevator car is visible, thereby indicating whether the car is at, above, or below the rescuer's level. Being able to see the counterweights (used to counterbalance the elevator car's weight), indicates that the car is a number of floors away, in a different direction. Counterweights at a lower floor indicate that the car is at a higher floor, and vice versa. Also, opening the top hatch of a car in the same elevator bank enables rescuers to look up and see the location of the stuck car.

Note: The law requires that these top hatches be secured so that they can be opened only from the top of the car. If this method is chosen, forcible-entry tools will have to be used to open the hatch from inside the elevator car.

Rescuers also might open the hoistway door at the entrance level and look up the shaft to locate the car. The hoistway door often can be opened with special elevator keys (Figure 11.3). The rescuers should look through the vision panel; and if they see the cable or counterweights moving, they should not open the door. A floor selector or leveler indicator in the machinery room also indicates the car's location.

Figure 11.3. *Special elevator keys are being used on this hoistway.* PHOTO BY AUTHOR

OCCUPANTS CAN HELP

Occupants' hollering and yelling, of course, probably is the easiest way to locate the affected car. Depending on their ages, how long they have been in the car, and how strong their vocal cords are, passengers stuck in elevators can lead the rescuers to them. Their yelling should not be taken lightly; many people are claustrophobic, and being confined in a small and often hot elevator car adds to their panic. After locating the car, rescuers should speak to the passengers and reassure them that the fire department is on the scene and in the process of rescuing them. This news should have a calming effect on the elevator's occupants.

An elevator can be stuck or stalled due to an electrical or mechanical problem. Sometimes these problems can be corrected by the elevator's occupants or by members of the rescue team.

After making verbal contact with the elevator's occupants and ascertaining that they are not hysterical or injured, rescuers may be able to relay instructions to them that will correct the problem. In many instances a broken electrical contact is all that is causing the elevator to stall. Rescuers should have one of the occupants push the car door to make sure that it is fully closed.

Rescue team members should check to see that all hoistway doors are in the closed position. Those closest to the stalled car should be checked first. The problem may have occurred just after the car left that floor.

Occupants should be instructed to press the *Door Open* button; if the car is at a hoistway opening, this may open the door at that floor. They can try pressing the first-floor button or the button for the floor at which the car terminates (intermediate banks of elevators can terminate at upper floors) in hopes of returning the car to its termination level. If they already have pressed the EMERGENCY STOP button, they first must deactivate it before any of the other buttons will work. Some elevators may have a special Fireman's Service: the use of a special key at the elevator termination point returns the elevator to that level (Figure 11.4).

Enlisting the services of an elevator mechanic is the easiest way to correct stalling problems, but a mechanic usually isn't at the scene when you need one. If all else fails, the occupants must be removed.

Figure 11.4. When using some types of special elevator keys, electrical power to the elevator must be shut down to protect the rescuer. PHOTO BY AUTHOR

Removing Passengers

When passengers must be removed, rescuers should do the following:

- First, shut down the power supplying the problem elevator. Two members should go to the machinery room; they should take with them communications to maintain contact with the members operating at the elevator. The machinery room may be located at the top or bottom of the elevator shaft or on a floor or two above the highest floor serviced by that bank of elevators. If keys are not available for the machinery room, forcible entry is necessary. Having two members present makes forcible entry easier and provides additional safety for operating in these machinery rooms.
- After gaining entry, shut down the power for the problem car. Regulations require that all switches be marked to indicate the cars they serve. If for any reason, the correct switch cannot be determined, all the switches must be shut down for safety reasons. One member always should be stationed at the power switch so that

*Figure 11.5A. Members using forcible entry tools to create opening for air bag. **B.** Air bag placed in opening is then inflated to force door open. (Note: A rabbit tool also can be used to force the door open.)* COURTESY J. CURRAN

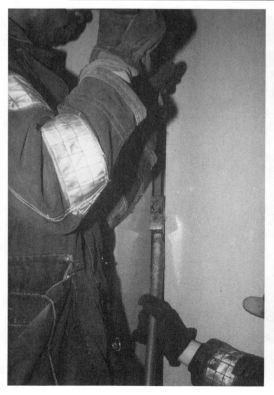

the power cannot be turned back on while members are working on the car. Even after operations have been completed, only a qualified elevator mechanic should restore the power. In the machinery room, members should be aware of possible openings in the floor used to ventilate the elevator shafts. They often are covered with a grating, which may be in a state of disrepair, or other materials used in place of the grating that do not provide the necessary protection or support.

- After shutting down the power in the machinery room, notify rescue team members at the problem elevator. They, in turn, usually can open the hoistway door with an elevator key or tool (Figure 11.5). (Normal work tools cannot be used to open the door.) If the occupants of the car can open the elevator door, it would be very helpful. A recent code change in the American Society of Mechanical Engineers standard includes a provision intended to prevent the occupants from opening the car door unless the car is within the "landing zone." This zone is considered to be the distance between the elevator landing and the car floor—a minimum of three inches above or below the floor landing to an allowable maximum of 18 inches above or below the floor landing. When the car is outside the landing zone, a restrictor bar prevents the car door from opening (Figure 11.6). To open doors that have restrictor bars, it is necessary to turn the spring-loaded restrictor bar at least one-quarter of an inch to clear the restrictor angles (Figures 11.6A, 6B). If the door can be opened and is within the landing zone, the occupants may be able to help open it.

    Another method for opening the door is to trip the locking device by poling; (Figure 11.7A) that is, a rescuer operating from an adjacent car or from a hoistway opening above uses a pole or hook to trip the locking device (Figure 11.7B) while a member stationed at the hoistway door on the landing pulls the door open. Rescuers must be familiar with the various types of doors (sliding, swing, center-opening), since each has its own locking device (Figure 11.8).

- The first option for removing passengers from elevator cars should be to bring them to the landing closest to the elevator car; it also is the easiest approach. At times, this method may not be possible (for example, in the case of cars that stop only at even or odd floors or if the stalled car is above the closest landing).

- Removing passengers through the top hatch opening and up a ladder to the hoistway opening above may be the only way out. When operating from above, a ladder is lowered to the roof of the elevator

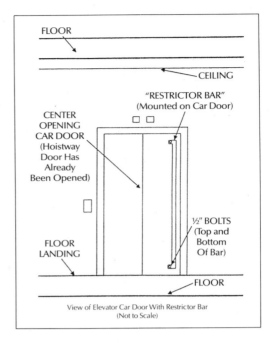

Figure 11.6A. Restrictor bar.

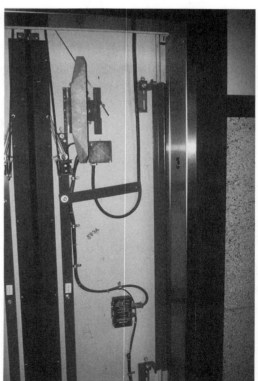

Figure 11.6B. Inside view of elevator door with restrictor bar. COURTESY M. MUNOZ

Figure 11.7A. *"Poling": Member using six-foot hook releases locking device of adjoining car.* ***B.*** *Adjoining elevator car door is opened without damage to either car.* PHOTOS BY AUTHOR

Figure 11.8. *Swing-type elevator commonly found in housing projects.* PHOTO BY AUTHOR

car, and two members climb down to the top of the car and open the hatch. Usually, a screwdriver or wrench is needed to open the hatch; often, the opening has been welded, bolted, or secured (to prevent illegal entry) and requires forcible entry.

If entry is through the hatch opening, a small ladder is placed down inside the elevator car, and one member goes down into the car and prepares the occupants for removal. Safety lines should be attached to rescue team members and elevator occupants during the removal process. Sufficient lighting during these operations provides additional safety. As is the rule when using ladders to remove victims, members must give the passengers as much help as they need.

- Cars stalled in *blind shaftways* present a much greater challenge to rescuers. Normally, access openings are required throughout the run of the shaft. These openings are spaced approximately every 36 feet, compared with hoistway openings, which are spaced every six to eight feet. The space requirements vary in municipalities, cities, towns, and states.

Additional safety measures are needed when removing passengers from a stalled car down to a hoistway landing. The opening into the shaftway from the bottom of the elevator car to the landing must be protected so that people will not fall into the shaftway. Also, the ladder from the landing to the elevator car must be securely fastened during removal operations.

During the New York City blackout in 1965, rescuers often had to breach the blind shaftway wall to reach trapped occupants. Breaching the shaftway wall requires locating the stalled car in the shaft and then locating the closest landing. This process may involve poking holes in the shaftway walls at different landings to get as close to the car as possible. Breaching may be necessary when an elevator's occupant is seriously injured or unconscious and can't be removed safely by other means.

In multicar hoistways, *a side-exit removal* is possible when another car in the hoistway can be brought level with the stalled car. Having an elevator key that controls the working elevator makes it a lot easier to line up the cars.

The side-exit panel in the car can be opened only with a key. If no key is available, forcible entry is required. Opening the side-exit panel on the stalled car from the shaftway side should not be a problem, as regulations require that a handle for opening this exit panel be provided on the shaftway side.

Once the side-exit panel of the operating car is opened and it's determined that the passengers can be removed, the power to both cars must be shut down. Although opening the side-exit panel breaks an electrical contact and eliminates the power, the switch in the machinery room also must be shut down for safety. Planks at least six to eight feet long, to span the opening between cars, are needed for car-to-car removal. They must be placed in both side-exit openings so that one rescuer can cross over into the stalled car. That member must have a safety line attached and after entering the stalled car, must attach a safety line to the occupants being removed.

After all the occupants have been safely removed, the side-exit panel of the operating car must be secured, and the power must be restored to the operating car through the machinery-room controls so that the car can be brought to a safe landing.

The most challenging of all elevator incidents is the one involving the removal of victims caught between the elevator and the hoistway door or hoistway shaft. Such situations could arise from accidents involving elevator maintenance personnel or games youngsters play with elevators. Victims' hands, feet, or other body parts, or even entire bodies, can be caught between the car and the hoistway (Figure 11.9).

Size-up Questions

Determining the answers to questions such as the following during size-up conducted as soon as the unit arrives on the scene, before implementing the game plan, establishes guidelines that can be used for large or small incidents.

Figure 11.9. *Placing an air bag between the elevator and the landing frees the victim's hand.*
COURTESY J. CURRAN

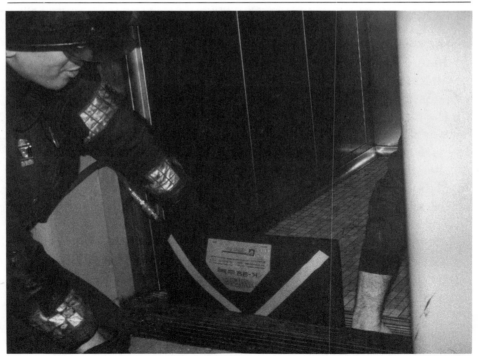

- Where in the elevator hoistway is the victim located?
- Is it a blind shaftway? If so, where are the required openings located?
- Can a hoistway door be opened near the victim's location to get a closer look at the situation? (In many instances the position of the victim's body blocks the hoistway door.)
- What is the most advantageous location from which to work?
- In multicar hoistways, is it possible to use an adjoining car to get to the victim?
- What is the victim's condition?
- Is he/she conscious and able to converse with rescuers and assist in the rescue?
- Is immediate first aid needed?
- Are medical personnel at the scene? If so, will they be able to assess the patient's condition? (*Note:* In operations involving deceased victims, speed is not as important as in operations involving a seriously injured victim.)
- How is the victim caught?
- In which direction was the car moving prior to the accident?
- Is the power to the elevator shut down?
- How much clearance is needed to free the victim?
- Can the elevator car be moved without further injuring the victim?

We responded to a report that a young lady had her leg trapped between the elevator car and landing. When we arrived, other units already on the scene were trying to calm her down. A language barrier had to be overcome before we could explain that the actions the rescue firefighter had to take would be somewhat embarrassing to her and to him. (She was about 17 or 18 years old, and he was in his twenties; the nature of the entrapment made it necessary for him to handle her leg and to move his hand up and down as he worked.) These actions, however, quickly separated her foot from the shoe, and she was completely free within minutes. After a quick "thank you," she disappeared without allowing herself to be assessed for injuries.

Experienced rescue firefighters sometimes can free victims without using tools. Determining whether this can be done is part of the size-up.

Another question might be, Is an elevator mechanic at the scene? Having a qualified mechanic present, as was indicated previously, can be very helpful. The mechanic can tell whether the elevator can be moved up or down, how much clearance can be gained, the parts of the car that can be removed to gain additional clearance, and how movement of the elevator can be controlled from the machinery room. Cooperation and coordination between rescuers and the elevator mechanic are essential.

Every elevator incident is unique. Members must draw on their experiences from previous incidents.

Safety Considerations

Safety for the rescuer and the victim are prime considerations. Safety in elevator incidents can be enhanced by doing the following:

- Securing safety lines to everyone working in hoistways
- Providing adequate lighting
- Having medical personnel stand by
- Leaving a member at the power shutdown location
- Providing communication between the machinery room and the incident location
- Using only trained and experienced rescue personnel—and only the number of members needed—instead of unnecessarily exposing members to dangers.

Elevator Fires

Fires can occur in the elevator car, in the hoistway, at the base of the hoistway shaft, and in the elevator machinery room. During fires in elevator shafts and assemblies, the main power supply must be shut down and all elevator cars must be checked for occupants before turning off the power (Figure 11.10).

Grease fires involving the cables or machinery on top of the elevator cars are very common. Rubbish accumulation at the base of the hoistway shaft often is the source of these nuisance fires.

The biggest problem caused by elevator fires is the smoke condition. A rubbish fire at the base of a hoistway shaft can send smoke billowing up the shaft, into the elevator car, and out on to the floors serviced by the shaft. These rubbish fires have resulted in mass evacuations of office buildings.

If an electrical fire occurs in the machinery room, caution is needed, and Class "C" extinguishers (for electrical equipment) must be used. A charged handline must be available to protect members and for fires that do not involve electricity.

Figure 11.10. Fire travels very rapidly up elevator shaft. COURTESY W. FUCHS

Fires within elevator cars themselves are not as common because there is very little inside the car to burn. Most of these fires are caused by carelessness and involve smoking or rubbish or are incendiary in nature.

Fire operations originating in apartments or offices can seriously affect elevator operations. Runoff water must not be permitted to enter the elevator hoistway shafts. Elevator components (electrical and mechanical) can be seriously affected by heat and water, which can cause erratic car movement. Salvage covers can be used to prevent runoff water from entering the shafts.

Walking up to the fire floor when the fire is below the tenth floor should be considered. It may take a little longer, but it ensures safe arrival. With today's emphasis on physical fitness, firefighters should not have any problems walking the 10 floors (Figure 11.11).

When using elevators, firefighters should always plan to get off at least two floors below the fire, and they should always have a means of communication when using elevators—be it the elevator car telephone or the intercom system. At least one member in the elevator car should have a handie-talkie. Should the car stall, the incident commander can be notified. An elevator car should not be used if there is a chance that the heat of the fire may have affected it. If possible, an elevator that services the floor involved, but that would not be affected by water or heat, should be used instead. An elevator should be tested by stopping at intervals of five to seven floors to be certain that the car is working properly. If problems are encountered, firefighters should stop the elevator at the next floor, get off, and use another elevator or walk up to the fire floor to gain access.

An elevator should never be overloaded; allowances must be made for the extra weight of turnout gear, hose, and tools. Forcible-entry tools may be needed if the car becomes stuck and the elevator car door must be forced open. The tools also can be used when the hoistway door is secured, preventing egress.

The location of the nearest stairway in relation to the elevator always should be determined; firefighters may have to use the stairs if conditions change rapidly.

Elevators equipped with the Fireman's Service allow fire personnel to control the elevator. In some systems, the activation of a smoke alarm or sprinkler automatically returns all elevator cars to their terminal points. Such systems usually are identified with special markings, indicating that they are for fire department personnel use only. The regulations, requirements, and restrictions that apply to this special fireman's service vary according to municipality, town, city, and state. Fire depart-

Figure 11.11. *In this building collapse, the single hoistway elevator remained standing despite the destruction inflicted on the building.* COURTESY J. IORIZZO

ment personnel must have a working knowledge of these systems, and they should hold orientation programs that involve fire personnel, building security, and maintenance and elevator employees so they can become familiar with the system's capabilities and limitations.

Training sessions on how to operate elevators during fires or emergencies should be a required part of every department's training program. Elevator companies will arrange for members to attend sessions that will give them a working knowledge of their elevators. Some elevator companies even provide training tapes. Housing complexes with numerous elevators often provide training drills for fire personnel and also may provide the special tools or keys needed to open their elevator doors. Building owners should provide rescue personnel with the necessary tools and knowledge to handle elevator incidents, and rescue companies should make sure they have them in their possession before an emergency arises.

12

AIR BAGS

COURTESY J. SKELSON

SINCE THEIR INTRODUCTION to the fire service in the 1970s, air bags have been used in many rescue operations. Their unique design makes them adaptable to rescue incidents and a vital resource for any rescue unit.

Air bags are constructed of Kevlar® Aramid® fiber reinforcement or of neoprene and steel wire. They are extremely tough, simple to operate, and require very little maintenance. Each bag has a single combination inlet/outlet valve used to inflate and deflate the air bag. The bags are resistant to many chemicals, gasoline and oil spills, punctures, scrapes, and cuts. They are capable of lifting, moving, and shifting objects weighing up to 72 tons, and they can bend steel bars, beams, or doors for forcible entry or to release entrapped victims.

Air bags have been used successfully in response to accidents involving automobiles, trucks, and other vehicles; rail or tank cars; elevators; heavy equipment; and, very frequently, also in building collapses. The mining, shipbuilding, pipeline-construction, stone-quarrying, rail-car repair, refining, utility-supply, and power-supply industries as well as the military use them in daily operations to lift, spread, shift, bend, force, or move.

As an interesting sideline, when the air bag first was introduced to the fire service, its name led some to mistake it for a type of bag that was designed to catch airborne people who had jumped from upper floors of buildings during a fire or other emergency. When new equipment or tools are introduced to a department, particularly specialized items carried by only selected units, the department should schedule a drill to familiarize members with the items' capabilities and limitations, especially if literature on the items is not immediately available to all department members.

Source of Power

The air bag gets its name from its source of power: air. The air used to inflate the bags can come from compressed air cylinders (SCBA), compressors, apparatus air systems, or a hand pump. The SCBA cylinder is the most commonly used air source (Figure 12.1). Adapters can be used to convert an alternative air supply (industrial compressors, for example) so that it will be compatible with the air-bag system.

Figure 12.1. Air supplied by an SCBA cylinder is the most common power source for air bags. PHOTO BY AUTHOR

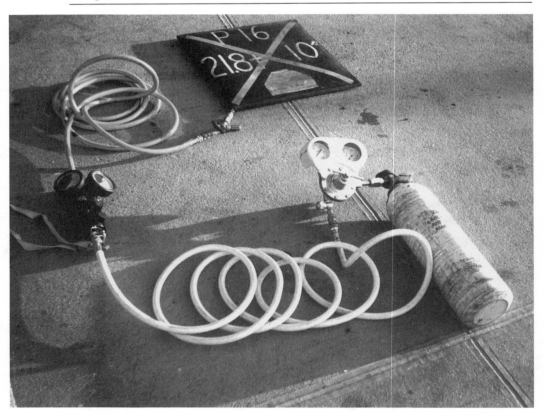

Types of Bags

There are high-, medium-, and low-pressure air bags. The low- and medium-pressure bags, operating at approximately seven to 14 psi, are used to lift, support, or move heavy objects. Overturned vehicles, tankers, rail cars, and the like can be uprighted or stabilized during rescue operations. Design features allow air bags to be used on all types of surfaces: ice, snow, sand, soft marshy ground, or rubble. Because of the lower pressure, they can be used against the thin skins of cars, buses, and airplanes without damaging them.

CYLINDRICAL BAGS

The cylindrical bags conform to all shapes that need to be lifted. Weighing as little as three pounds, they are capable of lifting 17 tons to a height of almost 70 inches. Manufactured in nylon and Kevlar® fabrics, they are highly resistant to oil and chemicals, tears, punctures, and heat; are extremely durable; and are available in special carrying cases that allow them to be transported to remote areas. For safety reasons, low- and medium-pressure bags should be used in pairs.

HIGH-PRESSURE BAGS

Available in a variety of sizes, high-pressure air bags range in size from six inches wide and six inches long to 36 inches wide and 36 inches long. They can lift from one ton to 75 tons to heights of between three inches and 20 inches, using a single bag. (By stacking two bags, one on top of the other, a height of 40 inches or higher can be reached.) The bags, constructed of neoprene, are reinforced with steel wire or Kevlar® Aramid® fiber. The Kevlar®-reinforced bags, which are approximately 40 percent lighter and much more flexible than those reinforced with steel wire, are easier to use in hard-to-reach places.

Each air bag (low-, medium-, and high-pressure) is labeled with its capacities for lifting height, tonnage, air volume (in cubic feet), and its operating pressure. Many rescue units use highly visible paint to mark this information on their bags (in addition to retaining the bags' labels, which usually are small and often become damaged) to serve as a reminder of the bag's limitations (Figure 12.2). The label also lists the safety guidelines to be followed during operations. A distinguishing mark such as an X or an O indicating the center of the bag is used to determine the bag's proper placement during operations; the mark is centered under the load (Figure 12.3).

Figure 12.2. *Marking bags with highly visible paint serves as a reminder of the bags' limitations.* PHOTO BY AUTHOR

Figure 12.3A. *The "x" on these bags is used for proper centering during operations.* **B.** *The center of the diamond is used as the centering point when using this type of bag.* PHOTOS BY AUTHOR

The high-pressure models—like the low-pressure bags—resist chemicals, gasoline, oil, punctures, scrapes, and cuts. Manufacturers offer complete systems, consisting of 10 bags of various sizes, or individual bags with the necessary components. If the department's budget prohibits the purchase of a complete set, the department can start with one or two bags and set a goal for eventually completing the entire set. A complete system includes air supply, air hoses, controller, and air bags.

SCBA-Supplied Air Bag

The compressed air cylinder (SCBA) is the most common means of supplying air. A high-pressure regulator reduces cylinder pressure down to the operating pressure. The regulator contains a low- and a high-

Figure 12.4. *Having sufficient air on hand is a vital part of the operational plan.* PHOTO BY AUTHOR

pressure gauge. The high-pressure gauge indicates the cylinder pressure; the low-pressure gauge is set for the operating pressure (usually between 120 and 150 psi, depending on the manufacturer's recommendations). A hose connected to the air-outlet connection is the supply line for the controller. This system—SCBA cylinder, regulator, and hoses—is portable, fast, and easy to operate, especially in confined areas.

The air bag label lists the cubic-foot capacity of air volume for complete inflation; an important part of the unit's operating plan must be to make sure that sufficient air is on hand (Figure 12.4); the amount needed depends on the number of bags, their capacities, and the number of lifts that may be required for a particular operation. A six-inch by six-inch bag requires .14 cubic feet of air, while a 36-inch by 36-inch bag requires 47 cubic feet. Obviously, larger bags and multibag operations require more air. Drilling and training with bags of various sizes and in different combinations develop confidence and assist members to learn the capabilities and limitations of the air bags and system.

Air Hoses

Air hoses come in various lengths and colors. It's extremely important to use different colored hoses in multibag operations. The operator of the controller usually cannot see the air bag being used. Instructions given to the operator by another member should be by color code to prevent confusion. Picture two bags positioned to lift a heavy structural member off a trapped victim. The bags are supplied by one controller and are at opposite ends of the structural member, out of sight of the operator. The member directing the operator relays, "Inflate the green hose," "Take it up easy on the red hose," and the like. Using colors to direct controller activity, particularly when the bags are completely out of the operator's line of sight, is simple and effective and cuts down the margin for human error. And, in air-bag operations, inflating the wrong hose could spell disaster. The bursting pressure of the air hoses is approximately 1,000 psi, well above the operating pressure. The hose has a male connection at one end and a quick-connect, locking safety coupling at the other.

Controller

The controller operates the air bag itself and is designed for single or dual capacity. A dual controller has a separate low-pressure gauge and safety-relief valve for each air outlet (Figure 12.5). Built-in safety relief

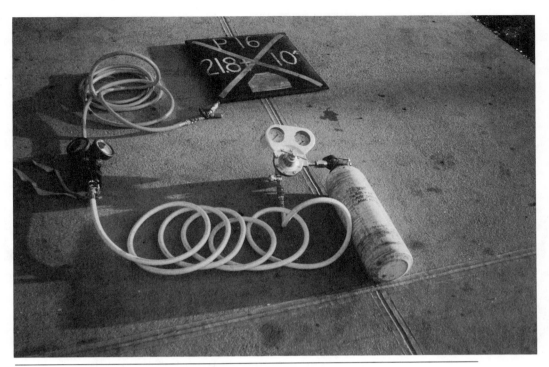

Figure 12.5. Bag, hoses, SCBA cylinder, regulator and controller. PHOTO BY AUTHOR

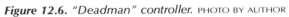

Figure 12.6. "Deadman" controller. PHOTO BY AUTHOR

valves prevent overpressurization of the bags. Various manufacturers offer simple one-quarter turn valves or "deadman" control buttons or toggle switches (Figure 12.6).

The *deadman control* plays an important part in any rescue operation. It is activated when the operator applies pressure and stops when pressure is released. (If the operator were to die while operating the control, activation would stop upon release of the button or switch—that's how the term *deadman* originated.) At one building collapse, members were in the process of operating an air bag to free a trapped victim when a secondary collapse occurred, burying the rescuer operating the controller and causing him to release pressure on the deadman control. The rescuer and the trapped victim were removed successfully, and the victim was spared any additional harm that could have occurred had the deadman button not been released.

CONTROLLER COMPONENTS

The components of the controller are:

- Air inlet valve: the hose coming from the air supply
- Operating gauges (There are two gauges on a dual controller)
- Control valve (Either deadman button/switch or one-quarter-turn valve lever)
- Safety relief valve with control knobs
- Air outlet connections (There are two on a dual controller)

Lifting Capability

The air bag's capability is directly related to its size. A six-inch by six-inch bag has a working surface of 4.74 inches by 4.74 inches, or 22.46 square inches. This area multiplied by the input pressure, say, 118 psi, gives the bag's lifting capability, which in this case is 2,650 pounds—the bag has the capability to lift more than a ton. A 36-inch by 36-inch bag can lift approximately 73.4 tons. Lifting heights range from as little as three inches to 20 inches, and stacking two bags of the same size doubles the height that can be reached.

The different shapes, sizes, and capabilities of the bags allow rescuers to adapt to the most difficult situations. The overall thickness of only one inch provides rescuers with a most versatile tool that can be used in the narrowest of openings with the greatest of flexibility.

Bag Sizes

SMALLER BAGS

The newest components to the air-bag system, the small-size bags are designed for limited-access lifting and spreading. These bags are especially helpful when the size of an opening makes it impossible to use a larger bag. In two collapse operations involving the freeing of a trapped victim, the six-inch by six-inch bag was the last piece of equipment used to conclude a successful operation. (The six-inch by six-inch bag weighs one pound and is capable of lifting 1.5 tons to a height of 3.4 inches, using 0.14 cubic feet of air.)

On another operation, the victim was still trapped after two hours of tireless efforts by rescuers. The lower part of one foot was securely wedged between sections of collapsed flooring. A rescue team member was able to dig out enough debris to make room for the six-inch by six-inch bag, which was placed in the small opening (Figure 12.7). With one inflation, the last remaining obstacle was overcome, and the victim freed. This experience helped in a similar situation, months later, when a victim's arm had to be freed before he could be removed from under eight feet of fallen building floors, walls, and furniture.

These small bags (six-inches by six-inches, six-inches by 12 inches, 10 inches by 10 inches) have been used successfully on numerous occasions to free victims entrapped in industrial and agricultural machinery (FIgure 12.8). How many times have we read about the worker whose hand or arm was trapped in a press or conveyor? These small bags certainly have become a vital part of a rescue team's arsenal.

Figure 12.7. LEFT: *A six-inch by six-inch and a 36-inch by 36-inch bag side by side.*
Figure 12.8. RIGHT: *Small bags (six inches × six inches, six inches × 12 inches, and 10 inches × 10 inches) have been successful on numerous occasions for freeing victims from industrial and agricultural machinery entrapments.* PHOTOS BY AUTHOR

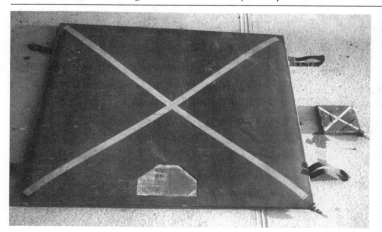

LARGEST BAG

At the other extreme is the largest high-pressure bag. Measuring 36 inches square, it can lift 73.4 tons to a height of 24 inches. Rescuing a bulldozer operator trapped under his machine required the use of two 36-inch by 36-inch bags. Placed on each end of the bulldozer, the bags were inflated in unison, allowing sufficient clearance to remove the operator.

Other Bag Rescue Operations

In another incident involving a high-pressure bag, a Doberman pinscher chasing a neighborhood cat became wedged under a large sanitation dumpster (Figure 12.9). A 17-ton bag was all that was needed to free the animal—but rescuers were more concerned about what the Doberman was going to do when freed than with the operation itself. Fortunately, the owner of the dog arrived at the scene prior to the freeing of the "victim."

Figure 12.9. *The owner of the victim (a Doberman pinscher) showed up just before rescuers freed the dog from under this sanitation dumpster.* COURTESY RESCUE COMPANY 2

It often amazes me how people can get involved in elevator situations the way they do. On an early spring night tour, our rescue company was called to an apartment house in response to a report that a man had his arm caught in an elevator. On the way to the scene, you try to picture the situation you are about to enter based on the information you have. My first thoughts led me to picture a repair man who accidentally had caught his arm while servicing an elevator.

When we arrived at the scene, we found two people holding a man whose arm was stuck in the diamond-shaped viewing glass of the hall-way elevator door; his feet were about two feet off the ground. The two good samaritans could do no more than provide support. The victim had broken the viewing panel glass in hopes of tripping the electrical con-nection to the elevator that had been stuck just above the door. As he put his arm through the opening, the elevator moved up, taking him with it. It had to be his lucky day, for as quickly as the elevator started up, it stopped again, leaving him hanging with his arm trapped between the elevator car and the shaft wall.

After setting up a chair for him to stand on, we set a game plan into operation. First, we shut down the electrical power to the elevator and left a member at the power source so it could not be accidentally turned on while members worked on the elevator. We gained access to the ele-vator shaft by the hallway door on the floor above. Entering the shaft, we found that we could operate better from the top of the elevator car itself. We then were able to use a 12-ton bag between the elevator car and the wall. We needed only inches to free the man's arm completely and to allow him to step down from the car. The man's "lucky stars" were still with him; after being freed, his arm showed only scrapes and bruises; he suffered no broken bones or open lacerations.

Train Incidents

The air-bag system has been put to very effective use in incidents involving trains, subway cars, or rail cars, be they derailments, crashes, or victim entrapments (Figure 12.10). Twelve- and 17-ton bags are ideal for lifting train wheel assemblies, axle assemblies, or a car from the wheel assembly. Actual operational procedures can vary, depending on how the train was manufactured.

Before setting up a departmental standard operating procedure for these types of incidents, a training session for the rescue teams and the agencies involved is a must (Figure 12.11). Understanding the system gives the rescue team an opportunity to incorporate safety guidelines

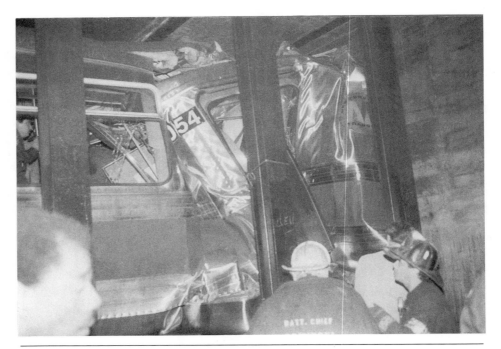

Figure 12.10. *Air bags are very effective in incidents involving trains, subway cars, and railcars.* COURTESY S. SPAK

Figure 12.11. *Training sessions are a must, especially when introducing new equipment to rescuers.* PHOTO BY AUTHOR

into its SOP. This process is extremely important, especially when electrified systems are involved.

Air bags often are used in accidents involving automobiles, trucks, and heavy equipment. As is the case with all rescue tools and equipment, the training and specialization of the rescue firefighter are important for maximizing the overall effectiveness of the air-bag system. Safety is an important part of every operation, and members should be dressed in protective gear to prevent injury.

Setting Up the System

Setting up the system using an SCBA cylinder involves the following steps:

- Connecting the regulator to the cylinder, making sure that the connection is tight and that the air source is opened slowly. The high-pressure gauge indicates the cylinder pressure.
- Setting the low-pressure gauge to the operating pressure (according to manufacturer's specifications). During operations, the high-pressure gauge must be checked to ensure that sufficient air is available. In extended operations, additional air should be readied.
- Making sure that all valves and controls are in the closed position to prevent accidental discharge of air.
- Attaching the regulator and then connecting the supply hose from the regulator to the controller, making sure that the locking couplings are secure (Figure 12.12). Opening the outlet valve on the regulator provides air flow to the controller.
- Checking to make sure that the connections on the controller are secure; then connecting a supply hose attached to an air bag to the outlet of the controller.
- Checking to verify that all connections are secure, and then positioning the bag under the load, with the air-inlet nipple pointing outward. Whenever possible, the hose should be attached to the air bag before placing the bag under the load. This method provides an additional safety factor by keeping the rescuer away from the load, limiting exposure time.
- Inflating the bag by using the control lever, button, or toggle switch on the controller.

Figure 12.12. *In-line valves should be placed on the bag prior to connecting the hose. Should the hose be used to supply another bag, the in-line valve ensures that the bag doesn't lose air.* PHOTO BY AUTHOR

Important Precautions

Bags should be inflated slowly to prevent shifting the load, and they should be inflated only as much as is needed, which depends on the incident itself. The controller's safety relief valve prevents overinflation of the bag.

Centering the bag under the load is extremely important. It provides a stable lift and prevents the bag from "popping out" from under the load. If a bag pops out, it can be extremely dangerous; the load loses its support, and the victim or rescuer could be injured.

When using two bags, the larger bag always should be placed on the bottom and the smaller bag centered on top. Two hoses of different colors should be used to prevent accidental inflation of the wrong bag, and the bottom bag always should be inflated first. If two bags of the same size are used, both should be centered under the load. The bottom bag should be inflated to its maximum capacity and then the top bag, to the height required. No more than two bags on top of each other should ever be used.

Using the largest size air bag possible yields the bags' maximum capabilities. Cribbing or shoring should be used when it is available, and using two bags gives additional height. Cribbing and shoring are used to gain height, support the bags, and protect the bags from objects or surfaces that could damage them; they also should be used to support the load that has been lifted to prevent shifting or to complete support. The rescuers' safety should always be the number one consideration.

Accessories

Accessories that increase the capabilities of the system are available. An *in-line safety relief* and *shut-off* attached to the air bag's nipple allows the air supply hose to be disconnected from the air bag after inflation. This device permits multibag operation, using the same supply line. Also available are *devices that can convert or adapt alternate air sources* to supply the system. Truck air-brake systems, low-pressure (90- to 125-psi) compressor systems, or other air-tool supply systems can be easily converted, using adapters. Because the air-bag system is versatile and components are readily available, rescuers can adapt these systems to handle almost any situation.

Maintenance

All system components should be kept clean (Figure 12.13). Doing this ensures dependable service. A stiff brush and mild soap and water should be used to clean the bag's surface. Checking for leaks can be done by inflating the bags to 30 psi and using the soap and water to check for bubbles. The bags' nipples should be checked for damage. Storing the bags with their nipples pointed upward and covered with protective caps keeps them in working condition. Hoses should be checked for damage to couplings and connections; locking couplings should be kept free of dirt and grime. Hoses also should be checked for cuts, cracks, nicks, and the like. The hose should be stored in the coiled position to prevent kinking. The regulator and controller should be inspected for cleanliness. The couplings must be free of damage, the fittings tight and not leaking, the valves and control knobs undamaged and in proper working order, the gauge coverings intact, the screws tight, and the gauges and safety relief valves in working order.

Regardless of the number of times the system is used, the entire system must be inspected at regular intervals that are incorporated into the department's tool and equipment maintenance schedule. This means connecting of all hoses, regulators, controllers, and bags.

It must be remembered that the scene of an operation is not the place or time to be checking tools or equipment. No matter how talented or specially trained the operator of the system is, the system can perform to its maximum only when it is in proper working order.

Figure 12.13. *Following the manufacturer's instructions ensures that the air bags will be properly maintained.* COURTESY PARATECH, INC.

13

UTILIZING THE RESCUE COMPANY

IT IS 3:30 A.M. on a cold winter night. The voice on the department radio in the cab of the rescue truck is firing out a preliminary report and giving instructions from the communications dispatcher in rapid-fire order. The tone of voice indicates a serious fire is in progress. The building involved is a four-story, occupied dwelling with heavy fire in the cellar and a smoke condition that has obscured the entire building from view. The first arriving unit reports that people on the upper floors are screaming for help and unable to exit the building. In addition, a heavy odor of illuminating gas is evident on the unit's arrival. Additional reports indicate that fire is extending out the rear of the building, is cutting off the fire escape as a means of egress, and is severely exposing an old church directly to the rear of the fire building.

After being notified of all existing conditions and his initial response assignment, the incident commander immediately requests additional help, including a task force of three pumpers, two ladder companies, an additional chief officer, and a rescue company. Standard operating procedures combined with good fire fighting tactics have guided the first arriving units to their initial assignments. The incident commander places his initial attack lines in position, and the ladder company members have laddered the building and are removing people from the upper floors (Figure 13.1). All units, except the rescue company, are on the scene and heavily engaged in their duties.

The incident commander has prioritized his immediate concerns and orders the rescue company on its arrival to assist in the search and rescue of victims from the upper floors of the fire building. His decision to use the rescue company this way is based on experience. Safely removing or rescuing the people trapped in the fire building is the primary responsibility in this incident. The incident commander decides to use the rescue company because of the magnitude of the fire and factors such as victims being trapped above the fire and the reflex time of the additional units (the time span between the incident commander's request for an additional unit and the time the unit arrives at the scene).

If the building involved was a boarded-up vacant building with the same fire conditions but presented no life hazard, would the incident commander order the rescue company to do the following:

- Determine the cause of the strong gas odor and shut it down from the street with its special gas shut-off keys?
- Place into operation special saws used for cutting the boarded plywood windows and doors (Figure 13.2)?
- Check the possible fire extension to the church directly in the rear of the fire building?
- Do all of the above?

In this case, the incident commander's knowledge of the rescue company's staffing, expertise, and response assignment combined with the fire conditions afforded him the luxury of issuing multiple assignments for one unit while awaiting additional help. The rescue company's staffing level of an officer and five firefighters would allow for three two-member teams capable of covering all three assignments. The officer and one member would check the exposure in the rear for extension, another team would operate the special saws on the boarded-up windows and doors, and a third team would work on shutting down the gas in the street. Not all rescue companies have six-member staffing (an officer and five firefighters); assignments should be made on the basis of individual staffing of the rescue company and knowledge of the members' expertise.

How should an incident commander use a rescue company? He/she bases decisions on factors such as the type and nature of the incident, the type of response assignment (initial or special called), the special equipment unique to the unit, and the expertise of the unit's members. Experience qualifies the incident commander to use the rescue company effectively (Figure 13.3).

Not every fire poses the same problems or is of the same magnitude. Incidents do not have to be complex, unique, or an odd job for an inci-

Figure 13.1. *Rescuers removing elderly victim from upper floor of rapidly spreading fire.* COURTESY W. FUCHS

Figure 13.2. *Rescue company member uses metal cutting power saw to open roll-down gates.* COURTESY W. FUCHS

Figure 13.3. *The services of numerous rescue companies were required to locate a fatally injured firefighter during this major fire and subsequent collapse.* COURTESY W. FUCHS

dent commander to put the rescue company to work. Fire has the potential to present a multitude of problems.

An incident commander about to declare the fire under control was notified that a heavy smoke condition reappeared above the original fire floor. Attempts to locate the source of the smoke condition were hindered by renovations that had taken place in the old building over the years. The incident commander did not use the rescue company initially

Figure 13.4. *Special heat-detecting camera being readied for use.* PHOTO BY AUTHOR

because of the first attack line's quick knockdown of the fire. He ordered the rescue company members to search for the source of the smoke condition with their special heat-detecting camera (Figure 13.4). Within minutes the special camera located the hidden pocket of fire in a recessed area of the ceiling; this fire pocket was causing the smoke condition, but it couldn't be detected by firefighters using conventional search methods. Being familiar with the rescue company's specialized equipment enabled the incident commander to use it with success.

If a serious fire in the cellar of a fireproof building with limited access needs immediate ventilation, the incident commander could choose to put the rescue company's combination drill and concrete breakers to work. Or the rescue company could be used to check for cracks or other visible signs of pending collapse.

Whether the task is ventilation, entry, search, or rescue during initial operations or using special tools and equipment at a large-scale operation, the incident commander's decision to employ the rescue company can have a major impact on the operation's outcome (Figure 13.5). The rescue company is an additional resource, regardless of the conditions.

Figure 13.5. This rescue company's cutting torch used in 1949 to gain entry to a fire in a paper storage warehouse is still used effectively in the 1990s. COURTESY E. HEAVEY

Emergency Responses

Emergency responses are a large part of every department's activities. In one of the largest fire departments in the world, emergency responses comprise 30 percent of the total annual responses. How does the incident commander effectively use the rescue company at emergencies?

Fire department units responded to a report of a gas odor in a multifamily dwelling. Such incidents usually are caused by a gas jet that wasn't turned off, defective stoves, or pilot lights that go out. Units at these responses usually find the leak's source fairly quickly by going from apartment to apartment to locate and rectify the problem. In this incident, however, the source of the leak remained elusive. Members checked and rechecked every apartment, stove, and gas appliance. A rescue company using a combustible detecting meter traced the source of the leak to a supply pipe in a wall and sealed off the pipe using a leak-and-plug kit (Figure 13.6). (*Note:* These items of special equipment were carried only by rescue companies in that department.) But, the incident

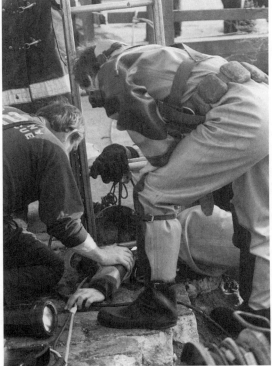

Figure 13.6. Rescue members enter the cellar to locate the gas shut-off during a major fire operation. COURTESY W. FUCHS

Figure 13.7. Rescuers combine scuba and confined space rescue talents. COURTESY S. SPAK

Figure 13.8. Rescue expertise is a must in collapse incidents. PHOTO S. SPAK

Figure 13.9. Rescuers save the life of a helicopter pilot. COURTESY S. SPAK

commander again prevented a minor incident from becoming a major one by using the special expertise and equipment of a rescue company.

One incident commander's solution to a major water leak was to call a rescue company to the scene. The expertise of scuba-equipped rescue members quickly resolved what could have been a serious situation (Figure 13.7).

A training session at which a rescue company displayed its newest equipment provided an incident commander with important knowledge, which he used at a serious incident the following week: a partial building collapse that trapped a worker under a large slab of concrete (Figure 13.8). The sections of broken concrete were too heavy for rescue workers to lift. A crane appeared to be the only way to move the concrete. Time was of the essence; the trapped worker had been seriously injured, and waiting for a crane could prove fatal. The incident commander ordered apparatus cleared from the immediate area and had the rescue company back its rig as close to the scene as possible. Having seen the A-frame, winch, and hoist attachment during the training session, he used this equipment to lift the concrete slabs and rescue the trapped worker.

Since rescue company responses are varied (Figure 13.9), the incident commander's familiarity with the rescue company's equipment and tools is the key that enables him/her to draw on these resources for major and minor incidents. Frequent drills and training sessions that include the incident commanders will provide that familiarity and enable them to incorporate these resources when formulating their plans.

The Incident Commander's Checklist for Rescue Companies

During the heat of the battle, incident commanders are required to run through a "mental" checklist of items they can use to formulate the strategies and implement the tactics needed to meet the formidable challenges of fire and rescue incidents. The incident commander's "computer" (the one under the helmet) should print the following out clearly and quickly (Figure 13.10):

- Is a rescue company responding? If so, has it arrived on the scene?
- How can the rescue company's members be used most effectively at this operation?
- What is the priority assignment for the rescue company?
- Is there a need for the special tools and equipment carried only by rescue companies?

Figure 13.10. *Rescue personnel use portable ladder from tower ladder bucket to reach trapped victim of airplane crash. Using a hydraulically operated spreading device lowered to them from the bucket, they successfully extricated the woman passenger.*
COURTESY CITY OF NEW YORK FIRE DEPARTMENT

- Will the rescue company have to carry out multiple assignments?
- Will the services of additional rescue companies be required? This could be the case for major collapses, train accidents or derailments, and other disasters. *Note:* Reflex time must be considered.
- How can the versatility of the rescue company be used most effectively? Will it perform multiple functions or be restricted to one assignment?
- *Proact* instead of react!

14

WATER RESCUE TEAMS

THE TERM *water rescue team* usually brings to mind a team composed of dive rescue members standing by waiting for an incident that requires their expertise in emergency situations. Most fire department budgets can't afford this luxury, and the state of the economy dictates that government, state, and local agencies get the biggest bang for their buck. The City of New York Fire Department solved this problem many years ago when it formed two "in-water firefighting units" following a number of disastrous pier fires that required the services of numerous firefighting units and lasted for days at a time. These teams were composed of members of two rescue companies who were trained in dive operations in addition to their normal firefighting and rescue responsibilities. The training involved taking firefighters (many of whom were scuba-qualified) from basic scuba certification to advanced rescue certification. Although the original intent was to provide an "in-water firefighting team" for the difficult pier fires, the availability of a number of suited-up trained rescue divers for responses has proven its worth on a number of occasions (Figure 14.1). Victims quickly and successfully were rescued from a van that crashed through a fence and was submerged in 15 feet of water in a creek in one incident, and a pilot successfully was rescued from a helicopter that crashed in a river in another.

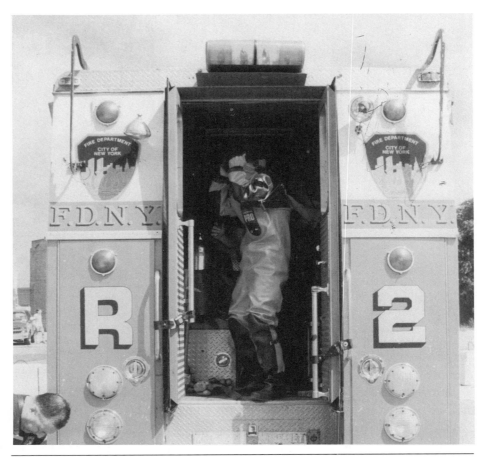

Figure 14.1. *Being able to "suit up" while en route to the scene saves valuable time and can mean the difference between a successful and unsuccessful rescue.* COURTESY RESCUE COMPANY 2

In-Water Team in Firefighting

Only a few months after their formation, the New York City "in-water firefighting teams" had the opportunity to prove their worth. A fire was spreading rapidly across the famous South Street Seaport pier, a 200-by 1,000-foot pier along a portion of New York City's East Side waterfront. The pier had been under renovation and was heavily loaded with building materials. A handline operated by a three-member in-water firefighting team extinguished the fire quickly and effectively, averting a possible conflagration. Operating from a 12-foot inflatable boat with a 15-horsepower motor, firefighters maneuvered between pilings. This maneuvering ability and the advantage of close-up extinguishment by operating handlines from the water are tactics not possible for land units (Figure 14.2).

In-Water Team Procedures

A high degree of coordination and teamwork are required to operate in-water handlines because footing is not as stable or firm in water as it is on land. Water reaction in the hose must be counterbalanced with the members' positions in the water. In-water firefighting teams generally use lines fed by land units. This procedure is safer than staffing lines fed by fireboats because the boat's propellers must be kept rotating to draft water, pump, and maintain position. This situation can be dangerous for in-water firefighting team members.

EQUIPMENT SETUP

Setting up equipment in water takes greater effort and more time than setting up on land. Moreover, major obstacles can impede in-water operations. A pier, for example, may be fully involved and necessitate that in-water firefighting teams enter the water and launch an attack from another area. Also, a particular area of the pier may be covered with a large number of pilings, making it necessary for the teams to change their courses of action.

Figure 14.2. In-water firefighting team operating handline at pier fire. COURTESY R. ATHANAS

Figure 14.3. Training session designed to improve operation procedures while diving from a raft. PHOTO BY AUTHOR

Versatility an Advantage

The services of these in-water teams also are needed in many incidents that do not involve fires. Their versatility has proven to be a unique advantage for rescue companies (Figure 14.3).

In one incident, a young man's car ran off an abandoned pier during the early hours of a foggy, rainy, fall morning. Units had been called to the scene after civilians reported seeing a car with lights on drive out on to the pier and the lights then disappear. No one could pinpoint the area of entry. The situation was complicated further when it was learned that the car went into the water at a location used as a local "graveyard" for abandoned autos. Divers, operating in dark, murky waters, known as "black water," used hand-by-hand touch and feel to locate the occupied vehicle. (*Note:* "Black water" is the term used to define water in which the diver expects poor visibility; in fact, any initial visibility, which sometimes is only six inches, is reduced to zero as the murky bottom is stirred by the diver's maneuvers.) In this incident, two dive teams worked in a relay system, taking turns diving and checking the area. The auto was located and the body removed. Had there been an eyewitness to report the incident sooner and to identify the point of entry more accurately, the victim might have survived.

Van Plunges in Creek

Such was the case only a few months later when a van carrying five youths went out of control, struck a fire alarm box, crashed through a cyclone fence, and plunged into 15 feet of cold, black water. Knocking over the alarm box opened the circuit in the communications office (dispatch center) and transmitted an alarm, which immediately was followed by phone calls reporting that a van had plunged into a creek. A rescue company with in-water capabilities was dispatched quickly. This rescue company truck carries two full sets of dry suits with AGA mask—a full facepiece designed for underwater use that gives a complete face seal with positive pressure of 120 to 140 ml—(Figure 14.4) regulators, tanks, weights, fins, knives, lights, and so on in the interior compartments of the apparatus. This unique factor allows two firefighters to suit up while en route to an incident, saving valuable time at the emergency scene.

Figure 14.4. *Diver protected with AGA positive-pressure mask and secured with a tethered line.* COURTESY RESCUE COMPANY 2

In-water operations involve removal as well as rescue tactics. A rescue operation depends on many procedures, but one of the major factors in determining the success of a rescue is the ability of the unit to put a searching diver in the water as rapidly as possible after arrival. An average sport diver takes 16 minutes or more to don the equipment. With proper training for rescue diving, this time can be greatly reduced. After drilling and testing, these dive teams have donned full dive equipment, while en route, in less than six minutes.

From shore, the submerged van could barely be seen in 12 to 15 feet of water. After the initial dive, debris kicked up from the bottom of the creek obscured the diver's view. The primary diver attached to a tethered line (guide line) entered the freezing water and made his way toward the van, which was about 25 feet from shore. The vehicle was lying on its driver's side. The roof had been crushed down, shattering the front windshield and negating entry through this very narrow opening. Finding all side doors locked, the primary diver tried the van's rear doors, which were unlocked. He attached his guide line to the handle to secure the location of the van and opened the door. Opening these two-by four-foot double doors took great effort because of water pressure and because the van's being on its side made it necessary to lift the doors. The first victim located was a 14-year-old girl, who was face down inside the van. With his arm securely around her waist, the diver surfaced, and other in-water team members helped to remove the girl from the water to the shoreline (Figure 14.5). Emergency medical service (EMS) personnel standing by began resuscitation and transported her to the hospital. This relay operation was executed four more times as divers, using the line attached to the door as a guide, dove back to the van for the other victims. During the rescue operation, the dive supervisor directed the second team of divers to search the area around the van for victims who might have been thrown from the vehicle. The divers were also instructed to attempt to unlock the van's front doors. After surveying the accident site, the diving team determined that no other victims were in the area and that victim removal through the van's rear door was the most practical method. It also was decided that only one diver would operate in and around the rear of the van, due to the limited size of the opening, poor visibility in the water, and the many obstacles around the van, such as discarded shopping carts, tires, bicycles, and the like. This strategy ensured that rescue efforts would be concentrated solely on the victim and not have to be expanded to include searching for a trapped or injured diver. After all the victims had been removed, a secondary search inside and around the van was made. The next day, a complete search of the entire area confirmed that all victims had been removed. One of the five youths

Figure 14.5. *This 14-year-old girl, a passenger in a van that ran off the road and became submerged in 12 to 15 feet of water, was rescued after being underwater for 20 minutes.* COURTESY W. FUCHS

removed from the van survived. The 14-year-old girl, who had been revived after having been clinically dead for more than a half hour, awoke from a coma; after spending 21 days in a hospital, she was released. Although she has some physical problems, her intelligence has returned to normal.

Survival Factors

Surviving such an ordeal depends on several factors that include the following: how long the person has been under water, the temperature (how cold) of the water, the condition of the water (degree of pollution), the age of the victim, the cause of "death" (drowning only—the absence of additional trauma), and the speed and level of expertise of the emergency treatment.

A victim is considered clinically dead when all vital signs (pulse, heartbeat, and respiration) cease. Four to six minutes after clinical death, biological death sets in.

Mammalian Diving Reflex

Although this 14-year-old girl had been under water for 20 to 25 minutes, her young age and the low temperature of the water were in her favor. Sudden face contact with cold water (below 70° F) sometimes touches off a primitive response called the "mammalian diving reflex." This complex series of body responses shuts off blood circulation to most parts of the body except the brain, heart, and lungs. Thus, what little oxygen remains in the blood gets transported to and pools in the brain where it is most needed. Even though very little oxygen may be in the blood, it can be enough, since the cooled brain requires much less oxygen than it normally does.

A large part of the young girl's survival depended, too, on the administering of proper and immediate first-aid and cold-water resuscitation by EMS personnel. Cardiopulmonary resuscitation (CPR) must be administered as soon as possible, as normally is the case. Care must be taken, however, not to warm the body extremities prematurely.

Emergency-Care Procedures

On-site emergency personnel should follow these procedures in cold-water rescues:

- Extend (spread-eagle) the victim's extremities away from the body.
- Warm the core of the body first (heart, lungs, and brain).
- Use warmed oxygen, if available, in the resuscitator yoke.
- Do not warm the extremities.
- Leave wet clothing on and shield the victim from the elements.
- Begin intravenous, if available.
- Transport cool. Do not use a heater in the ambulance.
- While en route notify the medical facility of the pending arrival of a cold-water drowning victim.
- Raise the core temperature to 76 degrees F before biological death is pronounced.
- Continue life support for one hour.

The brain is warming from the inside out. The section of the brain that controls the eyes has to be warmed before it can control pupil contractions. The pupils will not react to stimuli even though sufficient oxygen is being supplied to the brain. As noted, the efficient and coordinated effort of rescue personnel in the above incident was a major factor in the girl's survival. Some of the general procedures used in this incident can and should be applied to every in-water emergency rescue operation.

Operational Guidelines

Well before an alarm is received, the in-water team should establish a set of in-water operational guidelines, and every member of the dive team should attend an ongoing training program that has been adopted by the department. A good training program develops standard operating procedures that can be adapted to all incidents so that confusion and the resulting needless communicating and wasting of time will be eliminated.

Operational Procedures

On receipt of an alarm for an in-water emergency, the commanding officer of the dive team should do the following, if at all possible, while en route to the incident or as quickly as possible after arrival at the scene:

- Ascertain the correct location of the incident. In times of emergencies, incorrect streets or avenues often are given, causing delay. When every second counts, vague information must be made as accurate as possible.
- Obtain the location and phone number of emergency facilities such as recompression chambers (also called decompression and hyperbaric chambers), physicians, hospitals, Coast Guard stations, and so forth, near the dive location.
- Call Divers Alert Network (DAN), a 24-hour emergency service information center for dive accidents.
- Radio the dispatcher to contact someone at the scene, and do the following:

  1. Hold a witness (if one is present) at the scene until the dive team arrives. The information relative to the location of the submerged vehicle and time of entry can be invaluable.
  2. Have someone at the scene mark the in-water location of the incident so that valuable time is not wasted in the wrong area.

Use a landmark (tree or house, for example) that lines up with the location. At night, apparatus headlights and additional spotlights can illuminate the vehicle's point of entry.

3. Determine how many people are involved in the accident. Special call additional life support teams, if necessary. Have them at the scene for stabilization and transport before the victim is removed from the water.

4. Call additional fire department land units to the scene. Additional personnel are necessary for moving dive equipment to hard-to-reach areas. Extra equipment, ladders, lights, ropes, stretchers, first-aid supplies, and any other needed objects can be provided by additional ladder units.

5. Clear and section off the area close to the scene. Only essential personnel should be allowed entry.

6. Gather the following information concerning the water from the fire department's marine division, the dispatcher, the Coast Guard, and other sources. This information should include the following: the current's strength and direction, whether the tide is incoming or outgoing, the depth of the water in the general area, the amount of visibility that can be expected in the waters, and any other information pertinent to the operation.

• While en route to the emergency, the commanding officer should relay as much information as possible to the dive team and designate the primary dive and the backup teams.

If the response vehicle has the capabilities, the divers should don their gear while en route. This has proven to be the most distinct advantage in successful rescues.

All efforts should be directed to getting the primary dive team prepared. This dive team usually consists of two divers. A primary diver is assigned search and rescue and must be fully prepared for immediate water entry. A standby diver is also fully equipped and serves as backup to assist or help in any developing situation. Tenders are assigned to each diver to tether and communicate.

• Monitor the radio for any details or changes in the situation.

THE DIVING SUPERVISOR (Figure 14.6)

Upon arrival at the scene, the diving supervisor (who may also be the incident commander) should do the following:

Figure 14.6. *Dive commander discusses operational strategy with dive-team member.*
COURTESY RESCUE COMPANY 2

- Assume command
- Brief the primary dive team with all the information received from on-scene witnesses. A well-coordinated team effort is essential to a successful operation.
- Provide the divers with the plan of action. Each incident has unique circumstances, so no single set of rules can be used for every emergency.
- Have land personnel readied and prepared to treat and transport victims. Brief or review procedures if anyone is unsure. Be wary of the "Yeah, I know" responder.
- Verify that sufficient ambulances and personnel are available and that they are prepared for cold-water resuscitations.
- Inform the nearest hospital by radio of the type of emergency and the condition(s) of the victim or victims.
- Above all, ensure that every safety precaution is taken. A rescue attempt should not compromise the lives of the rescuers.
- At the conclusion of an operation, prepare a written report that includes all actions taken, the information received, the problems encountered, the results attained, and the pertinent occurrences. This report will be a permanent record of the incident and also can be used for critiquing the incident, to further improve the

unit's performance. The incident commander (dive commander) should be encouraged to take notes and record the times involved, which will be valuable to medical treatment personnel and for preparing the formal report.

Basics for Divers and Land Personnel

Certain basic procedures should be followed by land personnel, too, to ensure their safety and the successful outcome of an operation. It must be kept in mind, however, that unless it's an emergency and the possibility of a rescue exists, divers should not enter the water until all necessary and possible work preparation has been completed at the surface.

The following basics should be observed:

- All rescue operations are to be considered as life-and-death situations. All divers, therefore, must react on a rapid deployment rescue mode as per training.
- All diving should be conducted with a surface tender and line or with direct diver-to-surface communication. Line signals as well as a hard-wire communication system should be understood by all divers.

Diving-Operations Group

In search-and-rescue operations, four functions must be provided by the diving operations group:

DIVER

The tending line, minimum ⅜-inch diameter, is attached to the diver by a carabiner to a harness, a line around the diver's waist with a carabiner and quick-release knot (water knot), or a diver's bowline around the waist.

TENDER

The tender is the surface member of the diving team who works most closely with the diver on the bottom. At the start of the operation, the

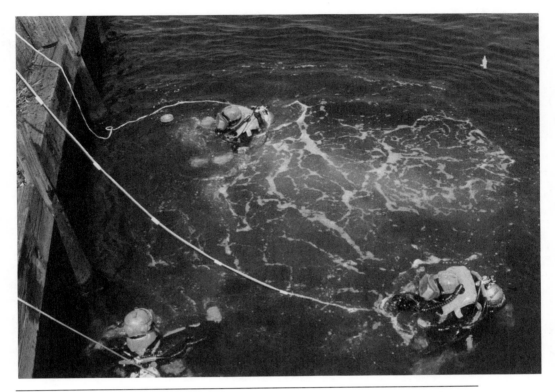

Figure 14.7. All divers are secured on tethered lines. COURTESY RESCUE COMPANY 2

tender checks the diver's equipment and topside air supply for proper functioning and assists the diver to suit up. Once the diver is in the water, the tender handles the tethered lines to eliminate slack or tension (Figure 14.7). The tender exchanges line-pull signals with the diver, serves as a backup to voice communication, and keeps the operation diving supervisor informed of the diver's depth and movements. The tender is also constantly alert for signs of an emergency.

STANDBY DIVER

The standby diver is assigned for backup or emergency assistance and should always be ready to enter the water immediately (Figure 14.8). The standby diver also should monitor the progress of the work as reported by the diver in the water so that if called upon for assistance, the standby diver will be mentally as well as physically prepared to respond.

Figure 14.8. *Divers in the standby mode. All divers should have backup divers in case an emergency arises.* COURTESY RESCUE COMPANY 2

OPERATION DIVING SUPERVISOR

For major diving operations, the operation diving supervisor generally will not enter the water. The supervisor's usual post is on the surface, in a position to direct tenders or standby divers. For simple and limited diving operations, the operation diving supervisor may also assume responsibilities as a diver and team leader.

When entering the water, divers should not leap; they should be lowered by a tending line to avoid contact with subsurface obstacles.

If a vehicle is involved, the diver locating it should tie either his tethered line or a marking line (a spare line) to the vehicle, which will provide a location marker that can be used as a guide line for the diver to follow. The primary diver should relay all information in relation to the vehicle to the dive team leader (the position of the vehicle, any obstructions, and so forth). Any sign that will assist in the rescue operation is important and should be used to brief other divers.

Dive team members standing by should be prepared to help remove the victim from the water to land. The time saved can make the difference in the victim's survival. If possible, the time the victim was removed should be recorded. It may help the medical personnel to determine the treatment needed.

Dive team members should not be distracted by the actions of land personnel, and other personnel should not be permitted to distract the dive team in any way. The divers should make mental notes of what they observed during the rescue or search and relay this information to the diving supervisor.

The victims should be removed as efficiently and quickly as possible. A victim never should be assumed to be dead. Time is of the essence here.

All dive supervisors, divers, and tenders should remain aware of decompression considerations during dives so that stage decompression is not necessary. Divers should always be equipped with a depth indicator and an appropriate means for recording dive time. Tenders also should record depths and dive time.

Cold-Water Rescues

The winter cold brings not only numerous firefighting problems, but also increases the need for body protection and training for rescuers engaged in cold-water incidents. The special training oftentimes can be the difference between a victim's surviving or dying.

In Fargo, North Dakota, a young boy completely recovered after being rescued from a lake in which he had been submerged for more than 30 minutes. A four-year-old boy spent 25 minutes in Lake Michigan before being rescued by Chicago firefighters. These types of incidents tax rescuers in the extreme.

A number of factors contribute to the success of cold-water rescues: the victims' ages; the water temperature; the mammalian diving reflex; the medical treatment victims receive; and, of course, the actions of trained rescuers. The latter is the most important.

Granted, unsuccessful rescues occur. Delayed notification to rescuers, lack of equipment, untrained rescuers, or a combination of the three may contribute to tragedies. These tragedies can be minimized by educating the public about the importance of promptly notifying the fire department not only of fires but also of such emergencies as ice rescues. An open house is an ideal place to demonstrate to the public the department's capabilities, equipment, and response duties. Most departments

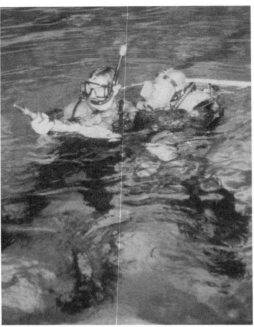

*Figure 14.9A. Training sessions using an inflatable raft. **B.** Training session in a pool.* PHOTOS BY AUTHOR

Figure 14.10. Dive-team commander discusses strategy with the incident commander. Knowing when not to put divers in the water is as important as knowing when to use them. COURTESY RESCUE COMPANY 2

hold open houses during Fire Prevention Week in October, which is the best time to educate the public about cold-water and ice rescues—when you are reviewing and preparing for such winter operations yourself.

Most departments have plans for high-rise fires, taxpayer fires, nursing-home fires, and haz-mat incidents, based on previous experiences. Even departments that have never been involved in a cold-water or ice rescue still should prepare for it. Planning should include reinforcing training with drill sessions, readying equipment, and formulating a game plan (Figure 14.9). The "computer" (under the helmet) must start outputting the information learned in training or from previous experience.

What can rescue units expect when they arrive at the scene? Most water-rescue operations involve some highly excited civilians and more than one victim. An additional victim may be a would-be rescuer— many courageous but untrained individuals become victims themselves while attempting to rescue others.

First arrivals should assess and evaluate the situation and gather information such as the number of victims; whether victims are visible or submerged; and, if submerged, the victims' locations or last known whereabouts. In the excitement of the moment, it may be impossible to obtain accurate information. The game plan then must be implemented based on the information at hand and adjusted later (Figure 14.10).

Common Procedures

The following are some common rescue procedures: *Reach, Throw, Tow, and Go; Reach, Throw, Row, and Go;* and *Reach, Throw, and Go.* The actions are similar for all techniques.

REACH

This approach involves considerations such as the following: Can the victim be rescued simply by the rescuer's reaching out and pulling him or her back to safety? Can the victim be pulled to safety by using an object such as a tree branch, pole, stick, rope, ladder, flotation device, or piece of clothing? Rescuers can be highly ingenious and improvise at these incidents.

Rescue teams carry such equipment as rescue ropes (Figure 14.11), buoy rings with ropes attached, rescue tubes, throw-rope bags, and life-line gun. These guns shoot a line to the vicinity of the victim, but the victim may not always be able to get to and hold on to the rope, line,

***Figure 14.11.** Practicing to use "throw ropes" to reach victims.* PHOTO BY AUTHOR

or rescue device. Victims may be physically injured from the accident or unresponsive, due to the effects of the cold water. The psychological reassurances that normally serve to calm victims may be ineffective during these operations. During the Air Florida plane crash in the Potomac River in Washington, D.C., survivors for this reason were unable to hold on to rescue lines lowered from helicopters.

TOW

If the victim is able to hold on to an object, rescuers on the shore can tow the victim back to shore. Rescuers must get as close as possible to the victim being towed; doing this speeds the operation without causing additional injuries.

ROW OR GO

If victims can't be reached by throw lines or ropes, rescuers must make their way out to them by rowing, or going in inflatable boats and rafts, aluminum boats, or rescue sleds. These vehicles are easily pushed across the ice surface. A line attached to the boat, sled, or raft enables other rescuers to pull the vehicle back to shore.

Removing the Victim

Two rescuers usually are required to remove the victim to the rescue vessel (Figure 14.12). These rescuers should be outfitted with specialized exposure or survival suits; such suits protect them from the extremely cold water and are designed to keep them buoyant. The suits can be easily donned over work clothes; they also have attachments for safety lines. All rescuers and victims must be secured with safety lines. The lines are not only for safety, but they can be used to tow victims or rescuers when necessary. During night operations, a safety line serves as a guideline to the rescuer should he/she become disabled or entangled.

The priority must be to get the victim out of the water and to the shore as quickly as possible (Figure 14.13). Medical personnel should be standing by to render the proper treatment. Special re-warming techniques are required to treat hypothermia, the lowering of body-core temperature usually associated with cold-water incidents.

Figure 14.12. Rescuers bring drowning victim to shore. COURTESY RESCUE COMPANY 2

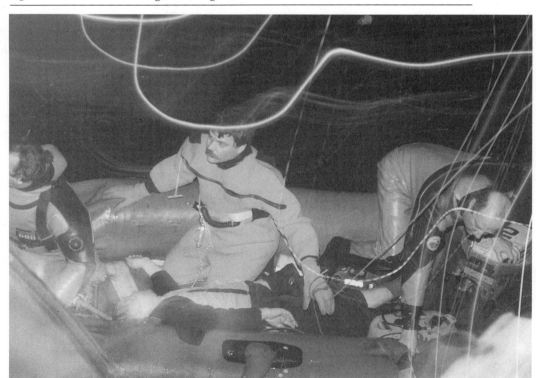

Figure 14.13. *This victim-removal technique involves lifting the victim in a stokes basket.*
PHOTO BY AUTHOR

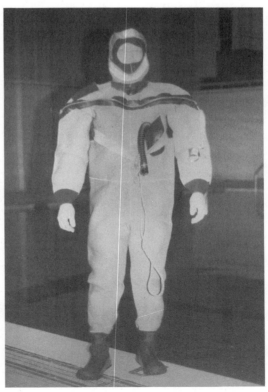

Figure 14.14. *Survival suits protect members from cold weather and/or water.* PHOTO BY AUTHOR

If the victim is submerged, only qualified divers trained in ice diving and rescue procedures should attempt a rescue. They must use dry suits with insulated underwear for protection against the cold water. Divers should always be tethered, and a backup diver must be suited up and standing by. Using improperly equipped or unqualified rescuers only compounds the problems encountered in these difficult operations (Figure 14.14). Safety begins with the rescuers.

Equipment

Equipment for these incidents can mean the newest on the market, if the department budget allows it, or the innovative use of more common materials or items. A number of years ago, members of a Massachusetts fire department had difficulty reaching a victim by ice sled during a rescue attempt. The rescue was a success, but the rescuers were not satisfied with the amount of time it took. The problem centered

Figure 14.15. *Diving gear, the motor, and the boat are prepared and readied on special response vehicle.* PHOTO BY AUTHOR

around the runners on the sled, which kept sinking into the "salt ice." The problem was solved with a little firefighter ingenuity. Using a surfboard and wearing a survival suit, the firefighters used two ice picks with strapped handles to test and cross the ice surface. The department since has refined the method and includes a number of these fiberglass surfboards in its rescue equipment inventory.

The department's inventory should reflect its response district and duties. Departments with numerous lakes, ponds, bays, inlets, or oceanfronts require a greater amount of specialized gear and equipment.

The growing popularity of such winter sports as skating, snowmobiling, and ice boating increases the potential for cold-water and ice rescues. Being prepared means having the equipment ready and in top operating condition (Figure 14.15). Medical technology often can save victims who have spent up to an hour submerged in freezing water. The efforts of skilled rescue teams, however, make it possible for these victims to make it to the hospital.

One point every rescuer must always keep in mind in any emergency operation, whether in water or on land, is to know when to quit.

Insufficient help, questionable equipment, and/or a too rapid current can make conditions unsafe for divers. Canceling an operation under such conditions doesn't signal defeat—it signifies common sense.

15

CONFINED SPACE

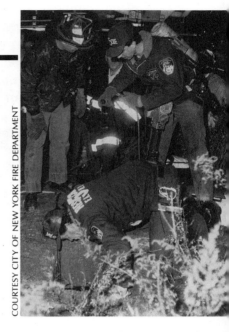

COURTESY CITY OF NEW YORK FIRE DEPARTMENT

DURING ONE EVENING DRILL, my lieutenant, who always preceded each drill with a "story" (whether he had actually taken part in the incident or was just "grandfathering" it), was in rare form. He had a habit of making multiple-alarm fires out of single-unit operations. The incident happened only a few months prior to my transfer into the rescue company; and a few of the members who worked at that scene were on duty this evening, which added credibility to the story. The story began with our unit's being called for a report of an unconscious worker in a manhole. When the unit arrived, not one victim, but two unconscious victims were at the base of a 12-foot-deep manhole. Questioning other workers at the scene revealed that one of the victims actually was a "rescuer" who had collapsed after entering the manhole. (According to the Occupational Safety and Health Administration [OSHA] and the National Institute for Occupational Safety and Health [NIOSH], rescuers account for more than 60 percent of confined-space fatalities.) How could this rescuer (now victim) have entered the manhole without an SCBA, in-line air supply, or testing the atmosphere? The rescuer had questioned the worker who had called for help. The worker, returning from a coffee run, discovered his coworker lying unconscious in the manhole. He didn't know what had happened, but he knew that the coworker, his friend, had been having medical problems. Consequently, he assumed that the victim had collapsed as a result of the problems. The rescuer made a gross mistake:

he assumed. In his eagerness to rescue the victim, the rescuer-victim had violated a number of safety procedures before entering the manhole and became a victim himself.

My unit often had drilled on confined-space rescue; members reviewed the standard operational procedure (SOP) and readied the necessary equipment. Using proper protective equipment, safety lines, air-extension lines, and teamwork, both victims were rescued without adding to the total of victims.

Incident Types

Over the years, a number of incidents involving confined spaces have gained national attention through the media and have been the subject of many fire service articles. Probably the most well-known is the rescue of an 18-month-old Texas girl who had been entrapped in an abandoned backyard well for 58 hours. Just a few months before that rescue, a boy was rescued from another well, this one in New Jersey. In Staten Island, New York, a man walking home across a vacant lot fell into a narrow opening (Figure 15.1); and had it not been for some inquisitive children, he could have become a grim statistic.

Figure 15.1. *A man walking across a vacant lot fell into this opening, which is being enlarged by rescuers.* COURTESY CITY OF NEW YORK FIRE DEPARTMENT

Construction workers falling into pits, shafts, and tunnels often become trapped in confined spaces. Manholes (utility, telephone, sewers, drains, and the like) provide a very likely setting for these types of incidents. Heavy machinery, too, can entrap workers in confined spaces. A worker cleaning a mixing machine fell into the machine. Before he could be freed, the victim *and* the machine had to be transported to a medical facility. Another worker fell into a stone crusher, and rescuers had all they could do to remove him from the narrowest of openings. Fortunately for both workers, the machines shut down, and neither was dismembered prior to being freed.

Possible Hazards

Our training teaches us that anytime we face limited means of entry or exit, we are dealing with a confined space. These spaces can be lacking in visibility and ventilation. They may be deficient in oxygen, or they may contain flammable gases or toxic substances. Furthermore, electrical, mechanical, and physical hazards also must be considered when undertaking confined-space rescue.

Answers Needed

When reviewing our "mental printout," the following questions need answers. The first one is, Why? What makes this incident different? Sometimes, the call is to help a conscious victim, which makes the incident relatively easy to handle. Percentage-wise, though, the majority of confined space incidents involve unconscious victims, and the operations usually are complex.

Only one thing is obvious: The victim needs assistance. Why did the victim fall? Was he/she overcome by one of the hazards of confined spaces? It is not uncommon for confined spaces to be lacking sufficient oxygen. The smaller the confined space, the easier for displacement of oxygen. These spaces should be checked with meters to assess the oxygen level of the atmosphere. (In this day and age, most departments have replaced canaries with sophisticated meters.) Self-contained breathing apparatus or air-extension systems should always be used, especially when testing with meters is not possible (Figure 15.2). Professionalism dictates that no chances be taken when doubt exists. Don't let machismo replace competent, clear, logical rescue thinking.

Figure 15.2. *Scuba-equipped rescuers operate at a confined-space incident.* COURTESY S. SPAK

Using a checklist helps ensure workers' safety. Strictly adhering to the guidelines enables workers to perform their tasks with a strong sense of confidence and understanding.

Lack of Checklist a Factor

Two utility workers might still be alive today if their company policy had included a similar checklist. We were standing by the scene of a reported gas leak, a common response for fire department units. A utility worker had entered a manhole to check for the source of the leak. The first engine company officer, following SOPs, had ordered a precautionary handline to be stretched. Truckies were checking the surrounding structures for evidence of the gas odor while rescue company members were setting up their meters to check for vapors in the structures closest to the manhole.

Additional utility workers arrived at the scene and were looking down into the manhole when an explosion occurred. Flames shot out of the manhole 40 feet into the air. Everyone within 15 feet of the explosion was knocked off his feet. Workers looking into the hole received the full impact of the blast, and two of them were killed instantly. The engine company extinguished the fire immediately. Efforts were instituted to rescue the worker in the manhole. The worker in the manhole had been in the process of eliminating the ignition sources when he accidentally triggered the explosion. The force of the explosion knocked him down, and he received less of the impact than his unfortunate coworkers. He was successfully rescued by firefighters.

An investigation revealed that safety precautions required at this type of incident were totally disregarded. Workers failed to ventilate the manhole, did not test for flammable vapors or gases, were not using explosionproof equipment or nonsparking tools, and did not have safety lines or harnesses. In their haste to correct the condition, they caused the explosion. Since that time, the company has established SOPs, guidelines, a checklist (Figure 15.3), and strict safety procedures to be followed by workers prior to working in manholes.

Supervisor Must Be Present

Supervisors must be at the scene. The existence of flammable and toxic gases in confined spaces must always be considered. Procedures for all rescue operations must include checking for the possible presence of dangerous gases with the appropriate meters. Oxygen, explosive, and hydrogen sulfide levels must be checked. Two gases commonly found in confined spaces are methane and hydrogen sulfide. Sufficient quantities of methane, a flammable gas, can exclude oxygen. Hydrogen sulfide, a colorless, highly toxic, flammable gas can cause immediate death in moderate concentrations.

CHECKLIST FOR CONFINED SPACE ENTRY

JOB LOCATION: _____   DATE: _____ WORKERS _____

PHASE I (PREPARATION)   YES   NO
1. Signs posted, no smoking, no open flames, safety zone..........   _____   _____
2. Piping into tank closed and tagged or locked out ....   _____   _____
3. Electrical disconnects locked out & tagged ....   _____   _____
4. Using nonsparking tools...   _____   _____
5. Using explosion-proof blowers in proper working order...........   _____   _____
6. Purged air sampled until below 10% LEL............................   _____   _____

PHASE II (SLUDGE IN WORK AREA)
1. Sludge drained and pumped out as much as possible, LEL below
   10%, O_2 level at least 19.5%, H_2S (hydrogen sulfide) 10 ppm or
   less. Continuous monitoring setup ....   _____   _____
2. SAFETY EQUIPMENT ON SITE:
   Resuscitator and first aid kit...   _____   _____
   Safety harness and lifeline...   _____   _____
   SCBA..   _____   _____
   Exploration-proof lights ....   _____   _____
   Wristlocks (if necessary)...   _____   _____
   Protective equipment for workers ....   _____   _____
3. Station man informed about job.......................................   _____   _____
4. Electrical equipment visually checked...............................   _____   _____
5. Safety check of man entering tank....................................   _____   _____

PHASE III (NO SLUDGE, DRY WORK AREA)
1. LEL below 10%, O_2 at least 19.5%.................................   _____   _____
   H_2S 10 ppm or less..   _____   _____
2. Tank monitored before entering.......................................   _____   _____
3. Station man informed about job.......................................   _____   _____
4. Two means of egress set up..   _____   _____
5. Welding equipment properly secured, fire extinguisher on site,
   cylinders outside of tank..   _____   _____
6. Monitoring being performed ....   _____   _____
7. Standby workers properly equipped ....   _____   _____

TELEPHONE NUMBER OF FIRE DEPARTMENT RESCUE COMPANY _____

SUPERVISOR'S SIGNATURE _____ TITLE _____

Figure 15.3. Sample checklist for confined space entry. ILLUSTRATION BY AUTHOR

Other Considerations

What are some other potential hazards in confined-space rescue? Attempting to gain *entry or egress* in itself is a hazard when confined spaces are encountered. These limited openings and spaces require rescuers to adapt and modify operating procedures and equipment (Figure 15.4). Working in and around physical obstructions in these spaces requires the special expertise and talents rescue firefighters develop through drilling, training sessions, and actual experience.

Obstructions can be in the form of ladders, piping, or conduits; they may result from an accident or a collapse within the confined space. When working in manholes, caution must be used since slippery or wet surfaces are common in below-grade areas. Manholes are used for electrical conduit, vaults, sewers, drain lines, telephone cables, gas lines,

Figure 15.4. Limited openings require rescuer to adapt and modify operating procedures and equipment. COURTESY CITY OF NEW YORK FIRE DEPARTMENT

Figure 15.5. Manholes are used for electrical conduits and vaults, sewers, drain lines, telephone cables, gas lines, pipelines—you name it, and it's probably in the manhole. PHOTO BY AUTHOR

pipelines—you name it, it's probably in the manhole (Figure 15.5). Rescue operational plans must include the possibility that these additional hazards could occur during operations in manholes. Whenever electrical or mechanical devices involved in confined-space rescue are shut down, a member should be stationed at the power source to ensure that the power remains off and is not restored until the incident commander approves it.

Usually, the larger the operation, the greater the *communications* problem. The additional number of portable radios and the need for information flow, orders, directions, strategies, and tactics by firefighters and officers increase the radio traffic. This problem would seem to be eliminated at confined-space rescue, where only one or two portable radios would be used, but what if the atmosphere is flammable or explosive? In such a case, explosionproof radio equipment is mandated. Alternate communications systems, such as sound-powered telephones, are effective in these situations. Voice and eye contact, hand or rope signals, slate boards, and chalkboards are other resources that can be substituted in place of explosionproof equipment. Safety and efficiency must be prime considerations. The contingency plan must include alternate communications systems.

In addition to using *nonsparking tools* and properly maintained and inspected electrical cords, lines, junction boxes, plugs, and receptacles, *explosionproof* lighting, blowers, and ventilation equipment must be part of the operating plans. Eliminating all sources of ignition and reducing the hazards for rescuers must be a major part of the safety plans. If intrinsically safe handlights aren't available, a trick that can be used is to securely tape the switch in the "on" position so that it can't accidentally be moved when operating in possibly flammable or explosive atmospheres. There is no such thing as being overly cautious when any of these hazards are possibilities.

Protective clothing should always include full turnouts, helmet, gloves, and SCBA or air-extension systems. At times, toxic atmospheres may dictate the use of "entry suits." These suits are designed to provide varying degrees of protection based on the type of material or hazard that may be encountered. Many of these suits provide in-line air connections that are incorporated into the suit's design.

Specialized Methods

Some unusual methods or modifications may be required for successful operations in confined spaces. The department's resources dictate the equipment that can best be used in these incidents. The fire service has drawn on modern technology for equipment that will help overcome many of the hazards and obstacles often encountered during these operations.

In operations where positive-pressure SCBA is required and entry or egress is being attempted by ladder through a narrow space or opening—necessitating the removal of the harness and cylinder from the rescuer's back—a coordinated *team effort* is a must. The rescuer in descent cautiously must work his/her way down the ladder as his/her partner at the opening positions the SCBA mask (Figure 15.6). The chances of accidentally pulling the mask off the descending rescuer is greatest during this phase of the operation. The rescuer crawling through a narrow opening may have to take the cylinder and harness off the back and push it to the front of the body as he/she progresses. Extreme care must be used.

Many of these areas are wet and slippery, adding to the hazards. A *lifeline* (safety line, tag line) always should be attached to the rescuer. It serves as the means back to the point of entry and the means by which fellow rescuers can locate the rescuer should he/she fall victim to the hazards of the confined space. The line also can be used for rope signals between rescuers.

As always should be the case for all types of operations, the scene of the confined-space incident is no place to test a procedure; frequent drills that simulate conditions in confined spaces should be part of the unit's drill schedule. All procedures, including mask operations, should be practiced.

Air Systems

In-line systems, extension-hose system, and *line mask* are terms for the same type of equipment that can be effectively used in confined-space rescue. They usually are designed to work as an in-line system only or as a combination system that includes a backup cylinder. OSHA requirements dictate that a five-minute escape bottle be used for all systems as a backup. Rescuers must be aware of the limited amount of air that will be supplied should the situation require the use of the backup cylinder. Air for in-line systems can be supplied from breathing apparatus, compressors, or cylinders of various sizes.

In preparing for worst-case scenario, the rescue unit should have an air system for the remote areas that cannot be reached by air hoses from apparatus or compressors. An in-line system employing a regulator attached to a 30-, 45-, or 60-minute bottle can be an effective alternative (Figure 15.7). These systems allow rescuers to get as close as possible to the incident scene without having to transport the larger, heavier cylinders. The system is not limited when used in conjunction with a second similar system. The system using quick-connect couplings allows for rapid changeover of air without any noticeable interruption in air supply. When compressors are used for air supply, possible interruption of air supply, mechanical or electrical failure, power supply, and refueling must be considered in the operational plan.

Having backup systems is another important safety measure. A worker lost his life when his in-line system, supplied from an electrically powered air compressor, lost its power source. Lacking the means for communications *and* a backup air supply system, the worker succumbed before rescuers could locate him. A number of SCBA systems allow a wearer to transfer air from his/her own cylinder to that of another member's. This capability could save the life of a trapped victim wearing an SCBA who could not be immediately freed. Of course, modifications to the SCBA, in-line, or air-extension systems should be done only by the manufacturers, and they must be NIOSH-approved.

Communications, a vital part of confined-space rescue operations, must be considered here, too. One of the safest and most effective meth-

Figure 15.6. *State-of-the-art SCBAs provide in-line capabilities, built-in communications, and buddy breathing systems.* COURTESY INTER-SPIRO COMPANY

Figure 15.7. *Rescuer wearing an in-line mask is lowered into a manhole during a training session.* PHOTO BY AUTHOR

ods of communication to use with in-line air supply systems is the sound-powered phone. Operating on sound power only, a possible ignition source (from nonexplosionproof radios, for example) is eliminated. The sound-powered line can be wrapped around the extension hoses of the in-line system and taped or secured by fastener ties to prevent the lines from snagging on objects. Headsets that employ a throat mike for transmitting easily fit under a helmet and are free from the mask facepiece. The nature of confined-space operations places greater emphasis on the need for a workable and safe communications system.

Victim Removal

Victim removal from confined spaces often challenges the ingenuity for which rescue firefighters are well-known. Removing victims by basket stretcher may prove ineffective. A number of alternative stretchers, designed specifically for use in confined space, are available. They have lifting points for vertical lifts that are very effective for narrow openings.

A "rescue pak" developed by a firefighter was specially designed for rescue operations. The pak is made of heavy canvas and is divided into three sections. The victim is placed on the center panel, and the two side panels are folded over the victim and secured with holding straps. Head and feet flaps are adjustable, and the pak can accommodate victims of any size. Specially designed webbing is interwoven into the pak and provides openings for six-foot hooks or wooden poles that can convert the pak into a stretcher for easier handling. A steel ring sewn in the top of the center panel is used for lifting in a vertical position. One of the main advantages of the canvas pack is that it can be maneuvered easily through narrow openings because it is not as rigid as stretchers and baskets.

Usually, a pulley system or block-and-tackle combination must be used to lift stretchers, baskets, or special rescue packs from manholes, vaults, narrow openings, and similar spaces. Many rescue units use prerigged systems (Figure 15.8). The pulley system or block-and-tackle must be secured to a stable object above the opening. Specially designed tripods are available for this type of work, or the lashing of two ladders can serve as an A-frame that holds the hauling system. One rescue-and-utility system available has a folding tripod, harnesses, safety belts, boson chairs, a chairlift rescue device, and overalls with a built-in harness for various types of rescue lifts. The hauling system uses core-sheathed or chemical-resistant ropes that raise or lower rescuers or victims, eliminating the need for block-and-tackle systems.

Figure 15.8. A tripod, a prerigged hauling system, and various lifting belts and harnesses.
PHOTO BY AUTHOR

An apparatus also can be used to provide an attachment point for a hauling system (Figure 15.9). The versatility of a tower ladder bucket (cherry picker, snorkels, aerialscopes, elevating platforms) can be deployed effectively when conditions allow for their placement at the incident scene. Aerial ladders also can be used effectively. In either case, it is mandatory that the manufacturer's recommended safety guidelines for using apparatus in these incidents are followed.

Figure 15.9. *An A-frame attached to an apparatus can be used for attaching hauling systems.* COURTESY P. FEATHERSTONE

Some heavy-rescue apparatus provide an A-frame setup as part of the rig. Portable poles attached to the body of the apparatus provide the attachment point necessary for hauling systems. Winches also can provide the hauling line for these systems.

Whether relying on high-tech or makeshift equipment born of firefighter ingenuity, following proper operating procedures and adhering to the checklist of safety items helps bring the incident to a safe conclusion.

Mental Checklist

The following is a *mental* checklist for confined space incidents:

- Oxygen-deficient atmosphere
- Flammable or explosive
- Poisonous or toxic
- Egress and entry limitations

- Electrical hazards
- Mechanical hazards
- Communications
- Visibility
- Ventilation
- Explosionproof equipment
- Eliminate ignition source
- Proper protective gear
- Safety lines
- Positive-pressure SCBA
- In-line system
- Backup system
- Test meters
- Removal equipment
- Stretchers, baskets, rescue pak
- Medical help
- Contingency plan

Using the checklist, guidelines, and proper protective measures ensures that the rescuer doesn't become the victim.

16

VEHICLE ACCIDENTS

COURTESY RESCUE COMPANY 2

EACH YEAR motor vehicle accidents account for almost half the nation's accidental deaths. In 1989, there were 12,800,000 vehicle accidents resulting in 46,900 motorist deaths. The number of accidents has steadily increased, as has fire department responses to these incidents. To deal with the increased responses and the more difficult extrication problems presented (the result of new technology in the automotive industry), rescue tools/equipment especially designed to meet the new challenges have become available.

Bringing a vehicle incident to a successful and safe conclusion, however, entails using this state-of-the-art equipment and a trained rescue team as part of an operational plan that encompasses all the appropriate and necessary steps, procedures, and actions. Rescue personnel must wear proper protective gear at all times during these types of operations, they must be specially trained to handle these emergencies, and they must maintain tools and equipment so that they are ready whenever needed. The apparatus taking the rescue personnel to the scene must be inspected and maintained on a regular basis as well. Training, however, is the most important element of a rescue company (Figure 16.1)—next to performing the actual operations at the rescue scene.

Figure 16.1. *Training is one of the keys to successful operations.* COURTESY J. REGAN

Training

Training for vehicle accidents can be accomplished in a number of ways. Drill sessions, for example, can begin with the basics such as the most frequent type of responses, expectations, and operating procedures and progress to the more advance staged multivehicle, mass-casualty drills involving various emergency agencies that would be participating in the more serious vehicle accidents. Training sessions should be as realistic as possible. Having a team member play the role of "victim" provides a better understanding of the victim's needs during a rescue operation. A hands-on training session with role players also provides rescuers with the knowledge needed when working around victims. During these sessions, a higher degree of concern for safety is more likely than in sessions involving an old abandoned vehicle without "victims," when the use of rescue tools and equipment is emphasized. Videotapes or slides presented during a classroom session prior to the actual hands-on training can provide some very informative preliminary advice and information that must be part of every team member's arsenal.

Calling on local car dealerships and their service managers also can be very helpful, especially with regard to the latest technological advances being incorporated into automobiles. Supplemental inflatable restraints (air bags) now are incorporated in many of the newer cars (Figure 16.2). Since September 1989, all new cars sold in the United States have been required to provide an automatic crash-protection system as standard equipment. In some cars, automatic safety belts fulfill this requirement. In other cars, air bags constitute the automatic crash protection. Manufacturers of these new cars provide literature that can assist rescue personnel, especially when dealing with cars equipped with air bags. These pamphlets and publications explain operating principles, some common problems associated with the bags, and any necessary actions that should be taken. The air bags inflate and restrain drivers and front passengers (in some models) after being activated by sensors during vehicle accidents. A training session with a car dealership service manager could be very helpful to the rescue teams likely to be working at vehicle accidents.

Like fires, no two vehicle accidents are alike; but frequent training sessions can help regardless of the circumstances encountered.

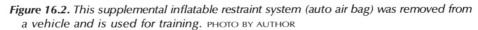

Figure 16.2. *This supplemental inflatable restraint system (auto air bag) was removed from a vehicle and is used for training.* PHOTO BY AUTHOR

At the Scene

Getting to the scene of the incident as quickly and safely as possible is the responsibility of the driver and the officer. During this phase of the operation, the officer must put his/her "computer" (the one under the helmet) into full gear. Taking the information at hand, the officer must begin putting the game plan into operation. First, he/she must choose a route that will get the unit to the scene without delay. The location or address received should be rechecked by asking the dispatcher to verify the information received. Imagine a response district that has an Oak Street, an Oak Avenue, an Oak Place, and an Oak Lane; having the correct location can save valuable minutes—an extremely important commodity, especially in life-and-death situations.

Drivers should be familiar with road and traffic conditions. An alternative response route should be part of every operational plan. Requesting traffic conditions from units already at the scene provides the information needed to reroute the response if doing this will get the team to the scene sooner. The officer should be aware of the type of response to the incident: Is, for example, an engine company, a ladder company, and a chief officer responding? An engine is a must, especially when dealing with gasoline, oil, hydraulic fluids, and extrication equipment. Having a charged handline at these accidents should be part of the operational plan also. Ladder companies provide additional manpower, tools, and an apparatus (elevated platform, aerial ladder) that can be used to gain access to above-grade elevated highways or areas where vehicles have gone over embankments and can be reached only by using portable ladders.

Positioning Apparatus

Determining where to place apparatus at vehicle accidents depends on a number of factors. If any of the vehicles are on fire, for example, the engine company must position its apparatus where it will be most effective in containing the fire. At nonfire accidents, the position of the engine company apparatus is determined by the location of the handline during extrication operations and gasoline and other fuel washdown considerations. When the aerial or elevated platforms of the ladder companies are not needed, the ladder companies can be positioned so that they do not block access for ambulances or rescue vehicles, but so that they protect other units at the scene from oncoming traffic (Figure 16.3).

Figure 16.3. *Positioning of apparatus at accidents must take into consideration controlling vehicle and pedestrian traffic.* COURTESY RESCUE COMPANY 2

Accident scenes can be dangerous areas for rescue personnel. It is good operating policy to have at least two units at the scene of every vehicle accident at which one unit is operating. Members of one unit should position apparatus and place flares to protect operating personnel. In addition, all apparatus' flashing and warning lights must be used to alert oncoming traffic. Many rescue workers have been seriously injured or killed as a result of rubberneckers trying to get a glimpse of the original accident.

Working Zones

An operational plan for vehicle accidents should include establishing working zones for rescue personnel. One of the working zones should be a "restricted" zone for rescue workers only. An adjacent "staging" zone should be for support personnel and staging of tools and equipment. These zones should be cordoned off to keep bystanders at a safe distance and out of the way of rescue personnel. When establishing zones, consideration must be given to leaving access routes open for rescue vehicles and ambulances that may not be at the scene. Rescue vehicles must be as close to the scene as possible; some rescue vehicles have equipment mounted in vehicle compartments with limited amounts of hoses and lines for their rescue tools.

The possible need for the apparatus-mounted winches, A-frames, and light towers (for illuminating the accident scene) should be considered before determining where the rescue vehicle will be positioned. The first arriving officer must assess the scene upon arrival to determine whether the number of responding units is sufficient to handle the incident. If there are any doubts regarding the capabilities of the initial response, the officer in charge should proact and call for additional help immediately.

Locating Victims

Among the first priorities in a vehicle incident are to determine the number of vehicles involved and to locate all victims. Questioning a victim with regard to the number of other possible victims sometimes adds to the confusion. Victims with head injuries or in shock may give rescuers the wrong information. Rescue personnel should look for some telltale signs or clues such as the presence of handbags, baby car seats, briefcases, and similar objects that indicate the number of people who may have been in the vehicle. It's common to find victims who have survived vehicle accidents wandering around some distance from the incident scene. Rescue personnel should check for victims not only in the vehicles, but also under and around the accident scene. People often are thrown distances from the vehicles after the impact. Small children have been found under dashboards and seats and covered by items such as boxes, blankets, pillows, rags, and such, that were in the vehicle prior to the accident. No location or area in or around the vehicle should be overlooked. Accounting for all possible victims and vehicles is an important part of the operational plan.

Hazards

While operating at the scene of a vehicle accident, rescue personnel must be aware of the hazards from oncoming traffic and from the accident scene itself. These dangers are posed by leaking fuel and other fluids, propane-operated vehicles, natural-gas-operated vehicles, the increased pressure in impact-absorbing mechanisms (BLEVE possibly) caused by the heat of a fire, and the new supplemental restraint systems (air bags). It's easy to see why at least one charged handline must be at the scene of a vehicle accident.

The incident commander must keep that "safety billboard" blinking brightly during the entire operation. As technological advances in the automotive field increase, so must the rescuers' awareness, knowledge, and training; they must keep pace with the challenges presented by modern day vehicle accidents. The incident commander's control of the operation can be the key ingredient to a successful vehicle rescue.

Stabilizing the Vehicles

To prevent further damage to the vehicles involved or further injury to victims or possible injury to rescue personnel, all vehicles involved in the accident must be stabilized (Figure 16.4). Rescuers can expect to find vehicles in almost any position—on their sides, on their roofs, and even on top of each other. Vehicles that remain upright after collisions also must be stabilized for extrication operations; otherwise, they easily can be moved during extrication operations, especially if they are on grades. All vehicles should be properly chocked: on the downhill side for vehicles sitting on a grade, and in both directions for vehicles sitting on level ground. Wheel chocks, cribbing, blocks, or other objects that can serve the purpose should be used. Chocking is in addition to ensuring that the transmission is in gear if it is manual or in the park position if it is automatic and that the emergency brake is set. Vehicles on their sides can be prevented from moving vertically by using cribbing, step chocks, air bags, jacks, ropes, chains, winches, nylon webbing, come-a-longs, or cables. The devices should be placed with extreme care so that they do not cause the vehicle to overturn. Obviously, tipping over a vehicle with victims still inside could prove disastrous for the victims, and possibly even for rescue personnel.

Gaining Access

Gaining access to victims can be as easy as opening a door or using an opening created by a broken window. The rule to remember here is, Always try before you pry. It never should be assumed that a door is jammed or can't be opened; it should be tried. Access usually can be gained by one of the doors that wasn't damaged in the accident.

If all the doors cannot be opened, access can be gained by breaking a window.

TEMPERED GLASS

An easy way to break a tempered glass window is to cover the window with duct or masking tape or contact or other adhesive-backed paper and strike the window in a lower corner with a pointed tool such as a pry axe, a halligan tool, or some other similar instrument (Figure 16.5). A spring-loaded center punch is the favorite tool of rescue team personnel for these situations. After breaking the window, a rescuer can pull the glass away while protecting the victims from flying glass by covering them with a turnout coat or a blanket.

LAMINATED GLASS

The laminated glass used in windshields has great resistance and requires special tools to remove it. It can be chopped with an axe or cut with an air-powered chisel, or a baling hook can be used to pull out sections from the windshield. Since it is bonded together with a sheet of opaque plastic and two sheets of glass, laminated glass presents less danger from flying glass than the tempered glass used in side and rear windows of passenger vehicles.

THE HATCHBACK

Hatchback-style vehicles provide another means for gaining access to victims: Forcible entry through the keyway and locking mechanism provides an opening to the locking device; inserting a screwdriver into this opening releases the hatchback. As soon as the victim can be reached, an emergency medical technician (EMT) or paramedic should stabilize the victim and conduct a primary survey to determine whether there are any life-threatening injuries (Figure 16.6).

During this phase of an extrication operation, the rescue operations officer can look over the entire situation and decide which removal procedures will be used and how the openings necessary to safely remove the victim(s) will be made. When the victim's injuries are serious, the operation must be accelerated. The rescuers' safety as well as the victim's safety must be prime considerations of the rescue operations officer during the removal process—even if it means summoning additional help or prolonging the operation.

Figure 16.4. *Vehicles must be stabilized during incidents.* COURTESY J. REGAN

Figure 16.5. *Windows should be covered with duct tape before they are broken.* PHOTO BY AUTHOR

Figure 16.6. *EMS personnel stabilize victim prior to removal.* COURTESY J. REGAN

Figure 16.7. *"Taking the top hinge."* PHOTO BY AUTHOR

DOORS

Hydraulic power equipment with spreaders, cutters, and rams are ideal tools for opening and removing doors, moving and removing seats, displacing steering columns and dashboards, cutting the A-, B-, or C-posts, and pulling and lifting.

The doors may have to be removed to provide an opening wide enough for the victim to be taken out of the vehicle.

To remove the door, rescuers should do the following:

- Position a hydraulic spreader on the top hinge of the door so that the door is pushed down and away from the vehicle (Figure 16.7).
- Continue opening the spreader until the hinge is broken and then do the same with the bottom hinge. Sometimes, it is possible to break both hinges without repositioning the tool. (*Note:* When hydraulic tools are not available, door openings can be made wider by using a come-a-long to bend the door back and out of the way.)

THE SIDE OF A VEHICLE

Should it be necessary to open the entire side of a vehicle, the B-post should be cut next, and if necessary, the spreaders used to force the rear door open. It now should be possible to push or bend the B-post and the rear door down and out of the way (Figure 16.8). A cut at the base

Figure 16.8. *Cutting the B-post.* PHOTO BY AUTHOR

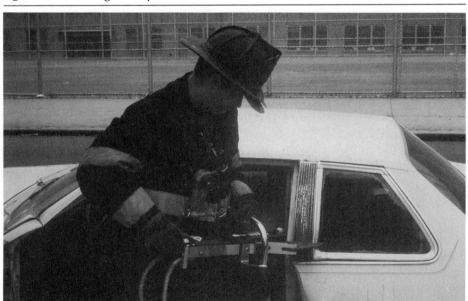

of the B-post allows the rear door and the B-post to be removed, providing a maximum opening.

THE REAR OF A TWO-DOOR VEHICLE

To remove a victim from the rear of a two-door vehicle, it may be necessary to create a "third" door from which the victim can be easily assessed, treated, and removed (Figure 16.9). This is done in the following manner:

- Removing the B-post
- Making a horizontal cut along the bottom of the B-post, parallel to the flooring of the vehicle
- Making two cuts in the side panel: one in the wheel well and the other straight down into the panel from the corner of the window.
- After making the three cuts, a spreader should be inserted while applying pressure directly on the underside of the roof where the B-post had been cut and on the side panel, pushing the side panel down and toward the rear of the vehicle. Opening the spreader as far as possible usually creates an opening large enough for rescuers to gain access for victim removal.

    Hydraulic rams also can be used to spread the side panel to create the "third door" opening. In this case, a short ram is used to start the opening and then extension pieces are added, or a larger ram is used to accomplish the same results as with a spreader.

THE ROOF

Removing the roof during extrications is a common rescue practice (Figure 16.10). Doing this early in an operation enables rescuers and medical personnel to gain earlier access and the maneuverability in the vehicle needed to treat and remove the victim.

Hydraulic rescue cutters are designed to cut easily through the metals and materials used in the roofs of today's vehicles. Removing the roof can be accomplished in the following manner:

- For safety reasons, the windshield should be removed before cutting begins.
- The A-post should be cut as low as possible to the dashboard to minimize the hazards created.
- Rescuers should be positioned so that they can hold the roof in

Figure 16.9. Making a "third door" opening will allow complete access to victims. COURTESY S. WILLIAMS

Figure 16.10. Completely removing the roof after cutting the A-, B-, and C-posts. PHOTO BY AUTHOR

place during the cutting operations and prevent its falling into the passenger compartment.

- The B- and C-posts should be cut; doing this will completely free the roof, which then could be lifted up and out of the victim(s)' way. In certain vehicle models, cutting the C-post or the rear section with an extra-wide post requires a number of cuts. They can be made by cutting a triangular section from the rear post which enables the cutters to be inserted deeper into the post for the final cutting (Figure 16.11).

Another method is to use the spreaders to crush the rear post, reducing the number of cuts necessary to free the rear post prior to removing the roof. When encountering problems with the wide rear post in a vehicle, the following simple method of folding the roof back and out of the way is recommended: After the A- and B-posts have been cut, cuts are made in the roof just ahead of the rear post on both sides of the vehicle. These cuts provide a pivot point for folding the roof back (Figure 16.12). In addition, cuts are made in the rear of the roof section on both sides; they serve as relief cuts and make folding the roof back easier. Although hydraulic spreaders, cutters, and rams have been used as examples of the tools necessary for these operations, other equipment such as air chisels, hacksaws, and power saws can be used. Obviously, using a high-speed saw will throw off sparks and create a very hazardous situation. If it's absolutely necessary to use this type of saw, a handline with a fog pattern must be used to minimize the dangers from the sparks flying around. The scene of a vehicle accident provides the ingredients (fuels, oil, and so on) that could spell disaster if a spark were to ignite one of these hazards accidentally. Using air chisels or hacksaws will not be as fast or effective, but again, if they are the only tools available, they can be used for cutting whenever necessary. The tool operator's proficiency can make up for the lack of some—but not all—of the tools' capabilities.

STEERING WHEELS AND DASHBOARDS

Many of the vehicle extrications involve the entrapment of victims against steering wheels and dashboards. The impact of the collision usually moves these vehicle components back into the victims; in rear-end collisions, the victim is pushed into them. Many "trapped" victims are freed by quick-thinking rescuers who apply the technique gained from previous experiences or training and simply activate the seat release. Many of the seat-release components survive the impact and operate as designed. Some power seats are wired so that disconnecting the wire

Figure 16.11. Cutting the C-post. PHOTO BY AUTHOR

Figure 16.12. "Flapping a roof." PHOTO BY AUTHOR

attachment allows the seat to be slid manually on the seat-assembly track. When necessary, hydraulic spreaders, rams, or other mechanical jacks, come-a-longs, and hand tools can be used to forcibly move the seat. Again, the rule is, Try before you pry.

At times, vehicle accidents require that the steering column and/or dashboard be displaced to free the trapped victims. A number of options can be used here, such as pushing, pulling, moving, or cutting. Steering wheels that tilt can be moved up and out of the way of the victims provided that the control lever hasn't been damaged. Rescuers should always try to move the tiltable steering wheel before cutting. Cutting sections of the steering wheel that are entrapping a victim can be done with hydraulic cutters, bolt cutters, or a saw. State-of-the-art equipment has made available small hydraulic cutters especially designed for use with steering wheels.

In instances where the steering wheel or column can't be moved or cut, pushing or pulling the column will be necessary. Pulling the steering column involves wrapping a chain or heavy-duty nylon webbing straps around the steering column and attaching another set of chains or straps to an anchor point, usually under the front of the vehicle. Another vehicle, tree, or pole can be considered when the accident vehicle has been damaged and provides no anchoring point. In that case, the two chains or straps are attached to a pulling device, usually a hydraulic spreader. The spreader is opened as wide as is necessary, and the chains or straps from the anchoring point and steering column are attached to the tips of the spreader, which have shackle attachments to use with the chains or straps. The spreader is then closed, causing the steering column to be pulled forward and away from the victim. It's advisable to use a come-a-long, air bags, or a vehicle-mounted winch, when available, to free the hydraulic spreader for other operations during an extrication. When using a winch, the cable must be in good condition and the winch and cable must be able to pull with the needed force to displace the steering wheel. Knowing the limitations and capabilities of all rescue tools and equipment must be part of the unit's resource inventory. Placing the tools/equipment on a base of cribbing helps to distribute the downward force and prevent the loss of lifting ability when the hood is crushed down or buckles under the force of the tool. Placing cribbing under the pulling chain or strap provides an upward pull on the column and often gives the clearance necessary for victim removal. When confronted with front-wheel-drive vehicles that have universal joints on the steering column at the floorboard level, it must be remembered that these columns can fail during displacement and cause additional injuries to the victim.

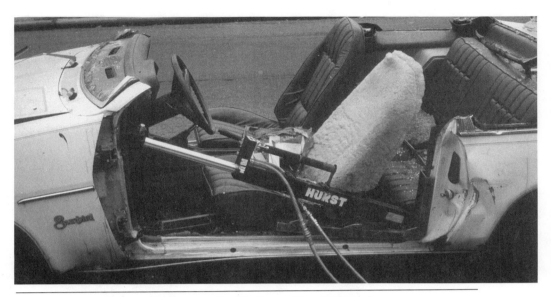

Figure 16.13. Using a spreading ram for displacing the dashboard. PHOTO BY AUTHOR

Another method for moving the steering column and/or dashboard is to "push." One of the disadvantages of pushing the steering column away from a victim is that this method requires rescuers to work close to the victim, making it possible that the victim would be further injured if the tool or equipment slips or springs back. Hydraulic rams with adapters designed to fit snugly under a steering column can be used to push the column up and away from a victim (Figure 16.13). The spreader also can be used for this purpose; extreme care must be used, however, when using this equipment near the victim.

Recent design changes and the use of unibody construction in automobile manufacturing, especially in General Motors products, have wrought changes in this type of victim entrapment. The new energy-absorbing steering column and its mounting design on the firewall cause the column to separate from the dashboard when severely impacted and is less of a problem during extrication. This type of entrapment makes it necessary to displace the dashboard to provide enough room for the victim to be released.

DASHBOARD

Displacing the dashboard may be accomplished by removing the vehicle's roof by flapping it back or by cutting the A-post first.

The next step is to cut at the base of the A-post where it meets the rocker panel. A ram then can be placed at the base of the B-post and high

Figure 16.14. *Using a hydraulic spreader, cutters, and a ram during a training session.*
PHOTO BY AUTHOR

on the A-post; by activating the ram in the position, the dashboard is pushed off and away from the victim. Placing wedges in the exposed area of the cut A-post reduces any movement of the dashboard when the rams are released. Removing the rams and maintaining the opening with the wedges provides sufficient space to treat and remove the victim(s). Special channel rails placed over the rocker panel and on the vehicle's bottom rail provides a base of support for the rams, formed at a 90-degree angle; the upright section rests against the bottom portion of the B-post, providing a secure base for the rams. This method of displacing the dashboard can be used on one or both sides of the vehicle, depending on the type and seriousness of the entrapment (Figure 16.14).

Removing the Victim

After rescuers have freed the victim, the victim must be packaged and prepared for removal. Packaging means dressing and bandaging where necessary, splinting fractures, and properly immobilizing a victim to prevent any further injury. These duties are the responsibilities of medical personnel. Many times, for safety considerations, the rescuers must help to remove the victim from the vehicle and to an ambulance. Many rescuers are trained as emergency medical technicians or paramedics and can give an extra (often times needed) hand to the medical personnel. When medical personnel are ready to remove and transfer the victim, rescuers assist them. Victims must be removed with extreme care. After having done so much hard work to free the victim, this is not the time to get careless.

An operation is not complete until the victim has been safely removed and transferred to a hospital or other health facility. An operation is not terminated until it has been determined that all victims have been accounted for, rescuers no longer are required, the equipment has been readied and placed back in service, and the apparatus and personnel are ready and available for the next response.

All the necessary reports can be done at quarters, and a critique of the operation can be held at the scene or back at quarters. Important, unusual, or vital tactics used in the operation should be discussed at the scene. Seeing the operation as it is being executed at the scene often leaves a lasting impression that can be used effectively at future operations. For rescue units, there never should be a job that's "just routine"; something should be learned at every response.

The skills rescuers employ at extrications can mean the difference between life and death, and they are the results of training and effectively using the lessons learned from previous incidents involving similar situations.

PART THREE

Operations and Planning

17

RESCUE OPERATIONAL PLAN

IT USUALLY STARTS about a week before the game. Coaches are sent to "scout" the opposing team. They note every offensive and defensive play of the game, paying special attention to the star players. Each player's strengths and weaknesses are important to the scouts. After the coaches discuss all the plays and players, they review the film of the game and formulate a game plan, developing offensive and defensive strategies devised to stop their opponents. Contingency plans are added so that adjustments can be made during the game. Many coaches can be seen on the sideline looking over their game plans during the competition (most of the plans protected by heavy plastic; experience has shown that coaches have rolled them into balls and have even used them as guided missiles during exciting moments).

Rescue operations, like athletic competitions, require a game plan, the rescue operational plan (ROP). Moreover, the rescue operations officer (ROO), like the coach, must develop a flexible and adjustable plan that will permit the ROO to implement any tactics necessary to control a rescue operation. The ROP can be used for collapses, explosions, fires, train accidents, derailments, plane crashes (Figure 17.1), ship disasters—just about any type of emergency or fire that needs a supervised, controlled, and coordinated rescue effort.

A rescue operations officer must have complete control over the rescue operation. If an incident commander is not on the scene when the

Figure 17.1. *An effective rescue operational plan proved very successful when rescuing these victims of a U.S. Air plane crash.* COURTESY CITY OF NEW YORK FIRE DEPARTMENT

operations officer arrives, the ROO must assume control so that the rescue operational plan can be implemented by an officer who has the knowledge and expertise to prevent the situation from exacerbating. The responsibility to direct, supervise, and control the rescue operation, however, does not relieve the ROO from coordinating and communicating with the incident commander once that officer has arrived on the scene. Furthermore, the ROO is responsible for maintaining communications with the incident commander and, more importantly, with the rescue teams.

During any rescue operation, a ROO's checklist of duties and responsibilities probably could fill one of Santa's mailbags, but his/her experience provides the "crutch" needed to handle the unique problems and difficulties as they present themselves. Rescue operations generally do not allow a ROO the luxury of using a standard form or guide to lead him through an operation, as usually is the case in fire prevention and inspection activities. Being able to make adjustments is the key element of an ROP.

The ROP at a Major Collapse

The following account illustrates how the rescue operations officer carried out and adjusted the ROP during a major building collapse.

It was a bright, sunny autumn day when the report of a building collapse was received in the dispatcher's office (Figure 17.2). What could

Figure 17.2. This collapse occurred on a bright, sunny fall afternoon in a busy commercial area. COURTESY J. IORIZZO

cause an apparently sturdy building to collapse when weather is not a factor? is a question that immediately comes to mind. Investigators have found that illegal or unauthorized renovations have caused many recent collapses; this was the case this particular day. -

A report received from a fire department unit near the scene at the time of the collapse conveyed the seriousness of the situation. Fire alarm dispatchers listened as a highly excited member told them he could not see the building—its remains—at the time. This report was the beginning of a major rescue operation that would have a sad ending for one family, but a very happy ending for many other families.

The first units arrived at the scene within minutes and were met by a number of people, some covered with dust and plaster, some bleeding, some hysterical, some dazed and wandering around, and others screaming for coworkers they believed were still in the pile of rubble that once had been a six-story brick-and-joist building.

The most apparent problem, one that demanded a great deal of effort and coordination, was determining how many victims were in the collapsed building.

INCIDENT COMMAND SYSTEM ESTABLISHED

Arriving shortly after the first units, the incident commander (a deputy chief) established a command post (Figure 17.3) and initiated an operation that would continue for more than eight and one-half hours, the time when the last survivor was brought out. As additional chiefs and companies arrived, rescue operation, victim control and coordination, safety, communications, water supply, and interagency liaison (police, utilities, buildings, and so on) responsibilities were assigned and the incident command system was established.

The ROO set the game plan into operation. The most accessible locations were addressed first. As additional rescue companies arrived, they were assigned to priority areas of the collapse. The three rescue companies were strategically positioned to provide a "six-sided" approach from above, below, and on each of the four sides.

COMMUNICATIONS

During a large-scale operation such as this one, all requests, directions, and information from rescue personnel must be channeled through the ROO, who is responsible for directing and controlling the operations of the rescue personnel. Since from 20 to 50 portable radios could be used during this size operation, rescue teams should use a sec-

Figure 17.3. "*The Command Post,*" *a vital element in the rescue operational plan.* COURTESY
CITY OF NEW YORK FIRE DEPARTMENT

ondary channel to avoid interference from other communications in effect. A company member with a portable radio set on the primary channel should be at the ROO's side and maintain communications with the command post.

THE RESCUE OPERATION

This building was unsupported by other buildings and had collapsed into a parking lot on the exposure 4 side, covering a number of automobiles. The remainder of the structure left standing was highly unstable. The initial rescue operation was centered on top of the pile of rubble where the upper floors had landed in a lean-to position, creating a few voids (Figure 17.4). Rescue personnel entered these voids and, using the "call-and-listen" approach (call: calling out and asking, "Is anyone in there?" or "Can you hear us?"; listen: listening for a response after calling), made contact with two trapped building occupants. The voices of these trapped workers gave the rescuers a point at which to direct their

Figure 17.4. A number of voids were created when the collapsed structure landed in a parking lot on top of several automobiles. COURTESY J. IORIZZO

Figure 17.5. Members standing by at the staging area. COURTESY RESCUE COMPANY 2

rescue efforts. The ROO assigned rescue team members to look for similar voids from the six sides. Their efforts proved negative.

The ROO notified the command post of the rescue effort in progress and requested that medical personnel be ready to treat the trapped civilians. Tools and equipment were readied for possible use. The ROO designated a staging area, separated from the command post staging areas, to be used for rescue equipment and personnel (Figure 17.5).

Since vibrations or unnecessary movements could cause the unstable remaining portion of the structure to collapse further, rescuers were ordered not to use power tools. Hand-by-hand digging and careful removal of the debris and other obstacles were substituted. Minimum manpower was committed to the collapse area to avoid creating an overload.

Debris was removed with extreme care while attempting to reach the trapped victims. Moving the wrong beam or supporting member could have caused an avalanche of debris. This portion of the rescue operation was one of its most critical aspects, and it reinforces the importance of using only experienced personnel in these situations. Another consideration in this incident was that the underground subway and rail traffic in the area could cause vibration. The agencies involved, therefore, were requested to shut down all service.

As rescuers were making their way to the trapped workers, the ROO placed additional rescue personnel on standby status, just outside the collapse zone (the area that could have been covered by the falling debris had the remaining portion of the structure collapsed). Too often, the safety of the rescuers is taken for granted.

THE "WHAT-IFS"

Another part of the checklist, the "what-ifs," could fill a page: What if the rescuer in this incident became trapped? (It happened in a large city; fortunately, the incident had a happy ending.) The ROO has to consider this possibility in situations such as this one. It is not unusual for the "computer" (the one under the helmet) to become overloaded during this type of operation. The most experienced rescue officer will tell you that after a "self-critique," it is usual to think of a number of items that might not have been on the "printout" that particular day.

After rescue personnel reached the trapped victims, stretchers were requested, the victims were secured in them, and a chain of rescuers safely removed the victims from the building within an hour after its collapse (Figure 17.6).

While the rescue efforts were taking place, the ROO asked the inci-

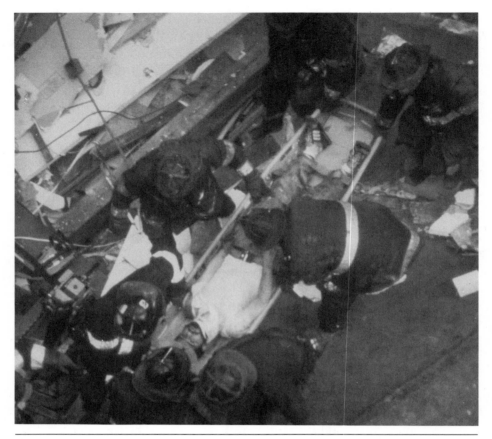

Figure 17.6. *One of the two trapped victims rescued early in the operation is removed by rescuers.* COURTESY J. IORIZZO

dent commander for any information that could determine whether other victims might be trapped under the huge pile of rubble.

While this information was being sought, a combination of actions dictated that all rescue personnel be ordered off the top of the pile while conditions were being assessed.

The continuous call-and-listen procedure now was getting negative results. The victims trapped for eight and one-half hours stated in an interview a few days later that she could hear the rescuers' voices and kept calling out for help. Her calls were muffled by the debris (as high as eight feet in some places) and, therefore, went unanswered. The portion of the structure remaining was highly unstable. The heroic initial rescue of the first two trapped building occupants had taken place under extremely dangerous conditions. The building commissioner reinforced the decision to remove the rescue personnel from the pile after he had completed a survey that indicated structural instability.

These factors made it necessary to reevaluate and reassess the rescue operational plan. Rescue personnel were directed to the operations staging area while the agencies involved were meeting at the incident command post to determine alternative actions to be incorporated into the game plan.

SEARCHING FOR THE OCCUPANTS

The number one priority was to ascertain the number and whereabouts of all occupants by interviewing the survivors at the scene and at hospitals (Figure 17.7); personnel, under instructions from the victim-control coordinator, visited the latter. The interviews, which also were conducted with eyewitnesses, occupants of adjacent buildings, and others familiar with the daily activities of the occupants, necessitated a team approach from all the agencies involved and revealed that 11 people were in the building when it collapsed. Two had been removed from the front section by an aerial platform; a few reached safety from the rear; and the others rode the collapse, like a giant slide, down into the parking lot. Two still were missing.

Another possibility lingered throughout the operation: People other than the occupants such as visitors, messengers, or passersby might have

Figure 17.7. *Interviewing witnesses or survivors of a collapse is an important part of every rescue operational plan.* COURTESY T. McCARTHY

been in the area when the collapse occurred. Fortunately, no "what-if" victims were found. The "computer" does not always have the answers to "what-if" questions.

Interviewing the survivors can yield information, such as the following, that is crucial for locating victims: Where were they last seen? On which floors did they work? In which portion of the building did they work: the front, middle, or rear? What is the normal layout of the floor? Did the occupants exhibit any particular work habits? Did they work at the same locations all the time, or did they travel to different areas of the building? Trying to determine where they were just before the collapse and then "reading" the collapse rubble to pinpoint location can be extremely helpful. This operation was proof positive that the interview method can be successful in building-collapse rescue operations.

In a proactive decision, the incident commander, taking reflex time into account, had requested that a crane be brought to the scene. In addition, a building department employee was asked to monitor the building continuously with a surveyor's transit for any signs of building movement that could forecast a collapse; a firefighter with a portable radio was stationed with him. Rescue workers were ordered from the pile, and the crane was placed in operation (Figure 17.8). As the crane worked, lights were set up, giving the impression that night had been turned into day. The crane operator later was highly praised for his surgeon-like removal of the remaining front portion of the structure.

The interacting agencies then focused on the next part of the game plan: separating the collapsed area into manageable sectors. The plan provided that every sector be commanded by a rescue team leader and that the ROO be the overall coordinator. Deploying the rescue teams in this manner gave the rescue effort greater coverage.

Before this part of the operational plan could be initiated, several checks had to be made, including making sure that utilities (gas, electric, and water) had been shut down. A gas accumulation sparked by a short circuit in an electrical line could have caused a disaster worse than the original collapse, and a cellar filled with water resulting from a broken water main could have undermined the remaining portion of the structure.

Rescue team leaders and members were briefed on the operational plan while the crane work continued. The areas to be searched were plotted according to the information gathered, an action that proved very helpful in locating one of the survivors. The building had collapsed and landed, surprisingly, almost in floor order; that is, as the floors settled in the parking lot, the roof area was on top of the pile, with the sixth, fifth, and fourth floor ceilings and floorings, respectively, under it.

Figure 17.8. *All workers were removed from the collapse pile while the crane was operating.* COURTESY J. IORIZZO

Figure 17.9. *Dogs, listening devices, and special imaging cameras were used during this phase of the operation.* COURTESY J. IORIZZO

The ROO, anticipating a prolonged rescue operation, requested additional manpower, sufficient equipment and tools for the staging area, standby medical personnel, and additional lighting.

After the crane operator had successfully picked apart the front wall of the structure—thereby removing the most imminent threat to safety—special "sniffing" dogs, listening devices, and a thermal-imaging camera were used in the search for the remaining victims, but they could not be found (Figure 17.9). Because of the possibility that the victims might still be alive, it was decided at the strategy conferences that hand-by-hand digging and debris removal would be continued and that heavy power equipment, such as bulldozers and backhoes, would not be used.

Where victim survival is an issue, the "Golden Day" (figure 17.10) is to collapse situations what the "Golden Hour" is to vehicle accidents. As a general rule, providing medical attention within 60 minutes (the "Golden Hour") of a serious vehicle accident increases the victim's chance for survival. Accordingly, data based on recent major earthquake disasters indicate that a collapse victim's chances for survival are much higher if the rescue occurs within 24 hours of the inception of the disaster—the "Golden Day."

Rescuers formed long human chains (Figure 17.11), and debris was passed by hand from man to man. Operations were stopped at one point so that some obstacles hindering the operation could be cut with a power saw. Prior to using the saws, the cutting area was carefully cleared and checked to ensure that a victim would not be cut accidentally. The power saws were used on a very limited basis, and shortly after, the hand-to-hand debris removal resumed.

LOCATING THE VICTIMS

After approximately four feet of debris had been removed from an area near the fallen section of the roof, rescuers located the body of the building owner. Before the rescuers became discouraged by this discovery, however, another rescuer called out to say that he had found a conscious victim.

Extreme care was taken at this point. Overanxious rescuers can further complicate the rescue effort if total control is not maintained. It is paramount, therefore, that the ROO take first-hand control. Surprisingly, the surviving victim was conscious and able to talk with rescuers as they worked to free her. She told them that she had been sitting at her desk on the sixth floor when suddenly everything seemed to break apart; she started to fall, as did the floor beneath her. She actually rode the collapsing floor down to the adjoining parking lot. Her fall was cushioned, and

Figure 17.10. TOP LEFT: "The Golden Day" for disaster victims. Earthquake victims have been successfully rescued after being trapped more than a week in collapsed structures. COURTESY S. SPAK

Figure 17.11. RIGHT: Rescuers forming "human chain" during rescue operation. COURTESY S. SPAK

Figure 17.12. One rescuer holds intravenous while paramedic prepares for removal, a victim who had been trapped for more than eight and one-half hours. COURTESY S. SPAK

she was completely protected by cases and boxes of novelty items stored on the upper floors. The boxes protected her from any crushing injuries that could have been caused by building components. A miracle, or just another "what if?"

The constant chit-chat between the victim and her rescuers helped to pass the time until she finally was freed; medical personnel evaluated and stabilized her condition during rescue operations (Figure 17.12). Hundreds of boxes had to be removed to free her body so that a stokes basket could be placed in the hole dug beneath her. Tired rescue workers became revitalized as they realized that this victim would survive; a feeling of satisfaction became evident throughout the entire rescue scene.

The possibility still existed, however, that the "what-if" victims could be in the pile. Another interagency conference was called to discuss this possibility. Plans were formulated for a rescue effort that was to continue for a number of days; it included using power equipment. Since no other occupants, workers, or civilians from this area had been reported missing during the eight and one-half hours since the collapse, it was believed that the continuing rescue effort would prove negative. It did.

The media were important factors in this effort. Television and radio coverage was extensive, and the rescuers believed that if someone who worked in or around the collapsed building area had not returned home, that person's family would have contacted someone at the scene.

A debriefing session was held after the operation had been completed, but the number of media representatives present made a critique and analysis of the operation impossible at that time. A few days later, the incident commander held a formal critique. Members reviewed photos, slides, and tapes. The consensus of the rescue personnel was that although the rescue effort was a success, the experience taught them many lessons that could enhance future rescue operations.

A Review

The following list is a review of some of the actions the rescue operations officer must take immediately to implement the operational procedures necessary to conclude a complex rescue operation successfully:

- Direct the rescue operation.
- Designate team leaders.
- Supervise, control, and direct the rescue team and the team leaders.

- Divide the collapse area into manageable sectors.
- Assess and evaluate the game plan as it progresses.
- Formulate a standby contingency plan.
- Gather information, collate it, and feed the "computer."
- Maintain communications—laterally with rescue teams/team leaders and upwards with the incident commander.
- Designate a staging area and the needed equipment, tools, and manpower.
- When time allows, run through the checklist: utilities, medical personnel, additional equipment and manpower, lighting, and so forth.
- Provide the incident commander with progress reports.

By following the protocol of a winning coach—implementing a game plan, using the "computer" under the helmet, following the strategies and tactics incorporated in the ROP, and making the needed adjustments—the ROO can improve the odds for a "win."

18

ROPE AND RIGGING

RESCUES IN THE FIRE SERVICE usually are accomplished by a variety of methods and means using equipment and tools that help to make the operation successful. Rope always has played an important part in the fire service, especially in rescue work. When I was a company officer, members of my unit used life-saving ropes in successful rescue operations on four occasions.

At one particular rescue operation, the firefighter involved was affectionately characterized by the media as a "Tarzan" in firefighter's gear. It was around six o'clock on a cold winter morning when my unit arrived and found a well-advanced fire in a crowded multiple dwelling, trapping a family of four above the fire. Street construction not only delayed the placing of a handline in position by the first engine company at the scene, but also completely blocked the street, preventing our ladder apparatus from entering it.

It didn't take long for the incident commander to realize that the problems being encountered could have disastrous results. Our company operating policy dictated that the roofman carry life-saving rope as part of his equipment (Figure 18.1). (The roofman's primary responsibility was to get to the roof of the fire building, provide all possible ventilation, and survey for possible fire extension and victims who might not be able to exit the building and needed assistance.)

269

Rope Rescues

At this fire, "Tarzan" quickly was notified that a number of victims were trapped at the third-floor front windows. Because the aerial apparatus was blocked from entering the street, another unit was dispatched back to the apparatus and ordered to bring portable ladders to the front of the building, in hope of reaching the trapped victims before the fire. "Tarzan," upon reaching the roof level, quickly deployed the life-saving rope and after being secured to the rope by his harness, was lowered from the roof to the third floor by the roofman of the second arriving ladder company. On reaching the window where the family was trapped, a highly excited parent quickly thrust two young children into his hands. As fire was venting out of the windows below him, he made a quick decision to swing across the face of the fire building to the adjoining building. A surprised occupant in this building, not realizing what was taking place, encountered outside her window a firefighter hanging on a rope handing her two small children. "Tarzan" had to make two more trips to rescue the mother and father.

Key Factors

As I look back on the four rope rescues performed by these courageous firefighters, I see that they had the following factors in common that were the key ingredients of these successful rescues:

- When entering the fire service, they received the identical rope-rescue training that was mandatory for all entry-level firefighters (Figure 18.2). The method taught and used at these rescues allowed the rescuers to use both hands to secure the victims when removing them and, if necessary, to signal the firefighter lowering the rescuer or to use a handie-talkie for communications. (The rescuers' hands were free because the rescuers were lowered by firefighters instead of having to lower themselves.)
- The rope used was specifically designed and intended for life-saving purposes only.
- The rope was well-maintained and always ready for immediate use when needed.
- All four members were in excellent physical condition and were maintaining their skills in rope rescue by participating in frequent drill sessions.

Figure 18.1. Roofman carries life-saving rope.
PHOTO BY AUTHOR

Figure 18.2. Rope-rescue training should start in recruit school. COURTESY J. DOWNEY

This combination of actions ultimately was responsible for the success of the four rescues.

Ropes

Although various types of rope are available, ropes made from synthetic fibers such as nylon and polyester are most frequently used for rescue work. Nylon and polyester are very durable and stronger, safer, and more abrasion-resistant than manila, cotton, or sisal.

Static rope (does not stretch) is preferred to dynamic rope in rescue work because dynamic rope stretches more and can create problems for rescuers when attempting to reach victims. Nylon static kermantle is the rope most commonly used for rescue work. Kermantle, which derives 75 percent of its strength from an inner core (kern) and 25 percent from a protective outer sheath (mantle), is less likely to spin and is more abrasion-resistant.

Laid rope, which consists of bundles of fibers (strands) twisted or laid around one another, on the other hand, tends to have resistance to abrasion, to be somewhat overdynamic, and to spin more than other ropes.

Other Uses

Rope has a variety of uses in the fire service. In addition to rescue work, it is used for search (Figure 18.3), tag and guide lines, securing

Figure 18.3. *Search rope is secured to door handle.* PHOTO BY AUTHOR

equipment, raising and lowering tools or equipment, lashing ladders, and even to provide fire lines. Rope used for life-saving purposes should be used exclusively for that purpose; it should be so marked and maintained as such. A log of the following information should be kept: the date the rope was placed in service; the type of rope; its length, diameter, and construction; its manufacturer; and its lot number. In addition, after each use, the following should be noted in the log: the date it was used, the reason for its use, and the user's comments after he/she has thoroughly inspected the rope. A complete history of the rope must be maintained. It is important to remember that the rope's life depends on many factors, including the following: care and maintenance, frequency of use, types of uses, types of friction devices used with it, methods for loading it, and degrees of exposure to heat, sunlight, or abrasives. Properly maintaining the rope ensures that it will perform flawlessly when it is needed.

Life Belts and Harnesses

Life belts and harnesses that can raise or lower a rescuer or victim during rescue operations come in a variety of sizes, shapes, and forms. Should these devices not be available, rope tied to anchoring devices on the roof or in a room—such as shafts, a staircase, or a radiator can be used. A number of fire departments train recruits to use these devices, and personal rescue harnesses and ropes are issued to the members on completion of training (Figure 18.4). These harnesses are designed to be worn at all times while on duty and can be used with a life-saving rope or a personal rope (usually 40 to 50 feet in length) carried in the member's turnout gear. It is used by members when it becomes necessary for a firefighter to reach a safe place when conditions warrant leaving the immediate area and the only means for doing this is to use the personal rope. This personal rope is intended for self-rescue only, not for victim rescue.

Protecting Rope

Rope should be protected from abrasions and cuts from sharp edges by using rope rollers, pads, edge protectors, roof rollers, or antichafing devices (Figure 18.5). A number of years ago, discarded fire hose was cut and modified to provide protection where needed when using rope over rocks or other sharp edges, or just to protect the rope from abrasions.

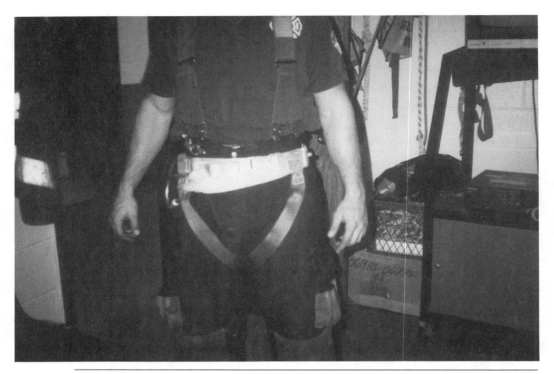

Figure 18.4. *Personal harness is worn by rescue firefighters.* COURTESY J. DOWNEY

Figure 18.5. *Life-saving rope with antichafing device (to protect rope). Life-saving belt is shown on left.* PHOTO BY AUTHOR

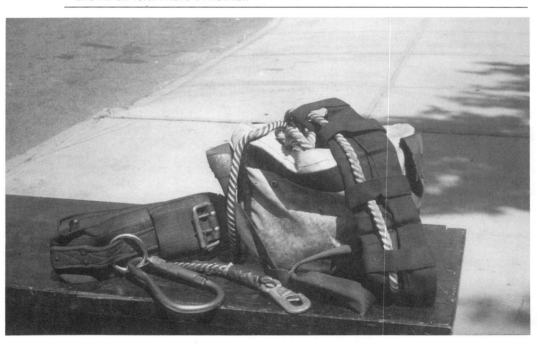

Storing rope in rope bags or rope packs helps to extend the rope's life.

Methods Used in the Fire Service

Rescue personnel throughout the country practice a number of rope-rescue methodologies. Some of these techniques involve the use of the following:

- A single rappel line anchored to a sturdy point, with the rope extended to the ground.
- A single rope with multiple anchors for the rope
- Two parallel and identical lines
- A two-member team using a single rope; one firefighter lowers the other (the method used by the City of New York Fire Department).

These methods are used also for mountain climbing and other activities that involve operating at heights that threaten safety.

Some inner-city urban departments use a system employing individual harnesses, life belts, and life-saving rope to effect rope rescues, while other departments use multirope approaches including belay systems, rappel racks, figure eights, anchoring systems, carabiners, and pulley systems to name a few of the numerous pieces of equipment and terms associated with rope rescue. Whether a rescue is accomplished by an individual single-rope technique or multiline systems employing a variety of equipment, one important factor generally determines whether the rescue is a success or failure: training that includes also frequent practice sessions to develop the confidence and skills needed to be successful in these operations (Figure 18.6).

NFPA 1983—*Life Safety Ropes, Harnesses and Hardware*, introduced in 1985, provides minimum specifications for rope-rescue equipment for the fire service; it certainly should be a source of reference for rope-rescue teams.

Rigging

In rescue work, lifting or moving heavy objects and/or weights requires hauling or lifting devices. A number of multipurpose systems for lifting and lowering personnel, equipment, and victims are available to the fire service (Figure 18.7). The hauling systems utilize single-line rope, galvanized or stainless-steel cable, or multiline, prerigged set-ups,

Figure 18.6. *The firehouse roof is being used as a practice tower during this training session.* PHOTO BY AUTHOR

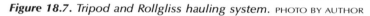

Figure 18.7. *Tripod and Rollgliss hauling system.* PHOTO BY AUTHOR

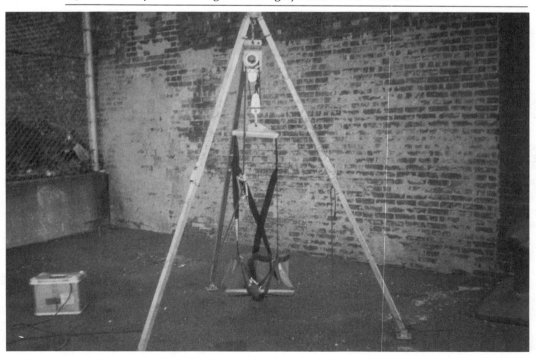

using the mechanical advantage afforded by the multiline systems. Using a block and tackle provides a gain in power. A block is a grooved pulley or sheave in a frame or shell that has a hook or strap by which it can be attached to another object. The tackle is an assemblage of ropes and pulleys arranged for hoisting, lowering, and pulling.

Simply put, the number of strands of rope between blocks determines the mechanical advantage ratio. For example, the effort required on a pull line to life a 600-pound object using double to double blocks is 150 pounds, or a four-to-one advantage (Figure 18.8). When using block and tackle, a number of safety precautions should be taken:

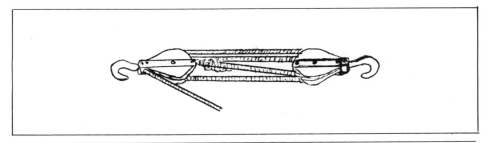

Figure 18.8. *Block and tackle with power gain of 4.* PHOTO BY AUTHOR

- The tackle must be the proper size rope/cable for the block.
- The blocks and sheaves/pulleys should be in good condition and kept clean and lubricated where necessary.
- The tackle must be the right size rope/cable capable of lifting the weight or object.
- When setting up the top block, the support holding this block must be strong enough to hold the load and the pull.
- A steady pull should be done simultaneously by everyone on the line; this can be accomplished by having only one "boss" to give the commands.
- Members pulling should be positioned so that they are not in a danger zone should any component of the system fail.
- Pulling should always be in a direct line with the sheaves/pulleys to prevent damage to the ropes, and cable (wire) never should be used on a block designed for rope.

Snatch blocks are used to change direction in pulling, which can be used for the safety of members, or to provide space for more members to pull on a line. Snatch blocks (Figure 18.9) can be opened on one side so that the rope can be inserted or looped onto the sheave without having

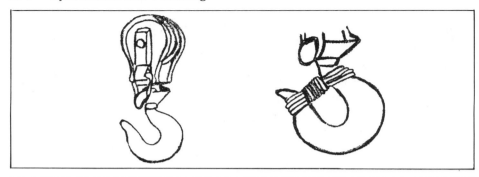

Figure 18.9. LEFT: *Snatch block.*
Figure 18.10. *"Mousing-a-hook": a safety measure that prevents a rope, sling, or cable from jumping out of the hook.* PHOTO BY AUTHOR

to thread the rope through the block. At times both ends of the rope will be secured and not capable of being threaded.

Mousing a hook (Figure 18.10) is a safety practice that prevents a cable, rope, or sling from jumping out of the hook. Sometimes it may be necessary to use prerigged hauling systems or block and tackle from ladders. This can be done by using a ladder on the floor above the working area, attaching the system to the ladder, and having the members at the working area raise or lower the victim, stretcher, or equipment.

Two ladders can be lashed together to produce an A-frame; (Figure 18.11) and by using slings properly supported at the point of attachment,

Figure 18.11. *Lashing the ladders together to make an A-frame.* PHOTO BY AUTHOR

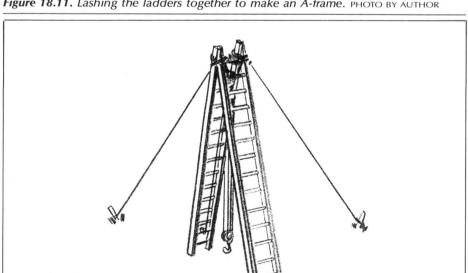

Figure 18.12. An A-frame on the apparatus provides a hauling system. PHOTO BY AUTHOR

Figure 18.13. The winch and A-frame on this apparatus were used to lift large sections of concrete from a collapsed stoop that killed two young girls. COURTESY CITY OF NEW YORK FIRE DEPARTMENT

the A-frame system can be used to raise or lower manpower/equipment into manholes or openings (Figure 18.12). Another method using a ladder is to secure the ladder by ropes to an apparatus or vehicle and set it over the area from which the lowering or raising will take place; it must be set at the proper angle (60 to 70 degrees), and the ladder must be able to support the load. When lifting loads, the load must be stable. This can be accomplished by placing the center of gravity directly below the main hook or point of attachment when lifting. The center of gravity of an object is the point at which the object balances. It is easy to see how important this balance can be when lifting or lowering a heavy object, equipment, or manpower.

Good safety rules require that rescuers know the center of gravity of the load; the weight of the load; and the capabilities and limitations of all the equipment being used, including hauling systems (Figure 18.13), lines, slings, chokers, and support points. A tag (guide) line should be attached, if needed; and the load lifted a few inches to check the load and rigging. Rescuers must start and stop slowly, maintain control of the load, and make sure to keep members clear of the lifting area.

19

TREATING VICTIMS

COURTESY W. FUCHS

WHEN FIREFIGHTERS THINK of treating victims, they usually envision seriously bleeding victims of a vehicle or construction accident, an industrial mishap, or a stabbing or shooting and tend to associate cervical collars, blood-pressure cuffs, dressing, gauze pads, bandages, splints, and resuscitators with these types of incidents. Many victims of rescue operations, however, need additional treatments, which are described in this section.

Psychological Treatment

Rescuers usually give victims psychological treatment before any other type of first-aid assistance, and it is given for a longer time for a complex rescue operation than for incidents such as the common type of vehicle accident, a shooting, or a stabbing.

Psychological treatment is extremely important. A seriously injured, conscious victim generally has two major concerns: How serious are my injuries; am I going to live? and How long will it take to free me? (The latter question usually is asked during vehicle accidents and major building collapses.)

Some victims who are unresponsive during a rescue operation later report that they could hear the rescuers but were unable to respond. I remember most vividly a young woman who tried to end her life by jumping from the fifth floor of a rehabilitation center. The building had a five-foot-high picket fence around the complex, and when she jumped she became impaled on the fence. The pickets penetrated her body in three locations: the thigh, the buttock, and the side just above the hip. The first rescuers at the scene were amazed to find that she not only was alive, but also conscious and pleading for help. As rescuers set up a backboard and provided stabilization shoring for the board, which became a temporary means of support for the duration of the rescue operation, the officer in charge began communicating with the victim. Psychological treatment was put to the extreme test in this incident (Figure 19.1). Here we had a young woman impaled on a picket fence, supported on a backboard at a height of five feet from ground level, and aware that the pickets went through her body. Her thoughts: Would she live through this ordeal? (Her original intentions were to do away with life.) How would the rescuers free her from this most unusual position?

We began by reassuring her that yes, she would be freed; time didn't seem important at that moment. As the rescue operation began, we again reassured her that she was in good hands—those of rescuers who had previous experience with this unusual type of incident. The rescue unit involved in this operation had responded to five impalement incidents in the past 12 months.

As rescuers began cutting two different areas of the picket fence with cutting torches (to expedite the operations), a protective fire blanket covered the victim to protect her from sparks from the torches. The officer explained the entire scenario to her step by step and the reasons for each action. He emphasized that it was important that she remain as still as possible to prevent additional injuries. By keeping up a conversation with the victim, the officer was able to take her attention away from her injuries and keep the focus instead on the rescue procedures underway.

By this time emergency medical service (EMS) personnel had arrived on the scene; but because of the victim's position on the fence, they initially were unable to perform a patient survey or assessment. Additional EMS personnel arrived and helped to ready the necessary equipment so that as soon as the victim was free and lowered to street level, they could start treating her without delay. They needed two stretchers, side by side, to support the victim and the separated section of picket fence that had been cut, but which was still embedded in her.

Cross-training between fire and EMS personnel makes it much easier to understand each other's problems and needs in situations like this (Figure 19.2). At this incident, because of the victim's position on the fence, rescuers had to free the victim and lower her before EMS person-

Figure 19.1. Psychological treatment of the victim continues throughout the entire rescue operation. COURTESY S. SPAK

Figure 19.2. Cross-training between emergency medical service and fire personnel makes it easier to understand each other's problems and needs. COURTESY S. SPAK

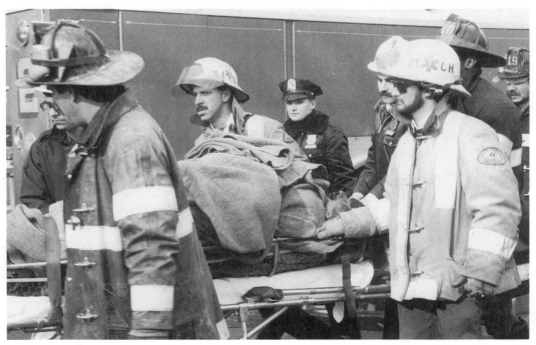

Figure 19.3. *Valuable time is saved when emergency medical service personnel are ready to treat victims.* COURTESY S. SPAK

nel could begin treatment; they saved valuable time by having their equipment prepared and being ready to start treatment as soon as the victim was lowered (Figure 19.3).

Rescuers helped EMS personnel remove the victim to an ambulance to ensure that the cut section of fence would be moved as gently as possible to prevent further injuries. The ambulance had to be stripped so that stretchers and personnel could accompany the victim to the hospital.

The incident commander ordered another officer to proceed to the hospital before the ambulance transporting the victim left the scene. As explained in the impalement chapter, this procedure serves a twofold purpose: to ensure that the entrance used to bring the victim into the hospital would be clear and large enough to accommodate two stretchers side by side and to inform the emergency room staff as to what had occurred and the nature of the victim's impalement and to ask them to set up a working area away from other patients. Because of the small size of the emergency room, this involved shifting patients and personnel. The officer suggested that all nonmedical personnel, including hospital volunteers, be removed from the areas through which the victim would have to pass and be treated. An older seasoned veteran of the emergency room objected to being told what was to be done in "her" emergency room. The officer tried to explain that the victim's appearance and condition might cause unnecessary disruptions if unneeded personnel were present.

The ambulance with the victim arrived at the hospital. The entrance had been made ready, and the two stretchers were wheeled into the emergency room. Two words faintly emanated from the area where the emergency room workers had been waiting for the victim. Some of the rescuers claim that the "holy s---" came from a seasoned veteran, while other rescuers were unsure about who had spoken them.

The sight of a victim with a picket fence sticking out of three areas of her body can be upsetting even to rescuers who have experienced numerous similar incidents. Credit has to be given to all the rescuers involved in this incident; they kept their cool, went about their business, and watched their tongues during the operation. This made it much easier for the officer in charge to render his psychological treatment.

The story had a happy ending. The pickets had not penetrated any vital organs, and the loss of blood was minimized due to the pickets' locations in the body. Paramedics and EMS personnel were on the scene quickly and stabilized the victim prior to her removal to the hospital. What at first appeared to be a possibly tragic incident turned into a success: After three weeks of hospitalization, the victim was well enough to go home.

Building Collapse

In another incident rescuers, after eight and one-half hours of digging, located a woman who survived being trapped in the rubble of a collapsed building. Psychological treatment for this victim took a different form than that given to the impaled woman.

This victim began a dialogue with rescuers, blaming the city mayor for the building collapse and her predicament. She cited his record on crime issues and environmental and social concerns—everything from daycare centers to multimillion dollar contracts with out-of-town vendors. While she expressed her opinions, rescuers had cleared enough debris to allow paramedics to make an initial patient survey. Rescue work was halted so that paramedics could start an intravenous line, which began her treatment for dehydration and shock (Figure 19.4).

At this point rescuers realized that the woman's injuries were more serious than they at first seemed. Paramedics were able to stop her from

Figure 19.4. *Rescuers await final preparations for removing the victim of a collapse as para-medics give treatment.* COURTESY S. SPAK

Figure 19.5. *Proper and prompt treatment of victims is essential to their survival.* COURTESY
J. REGAN

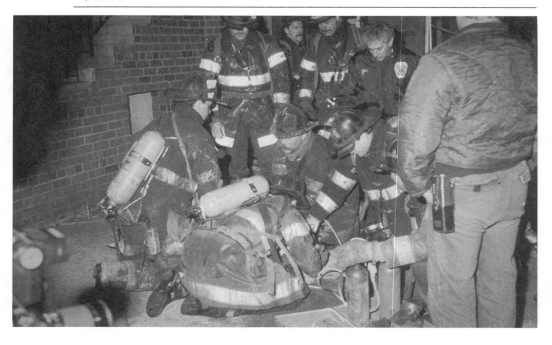

talking about the mayor long enough to get information about her injuries. They then informed the rescuers that the victim probably was suffering from crush injuries resulting from the mounds of debris that had been on top of her. Crush injuries can lead to crush syndrome (discussed below), defined as shock and renal failure following a severe crushing injury that results in trauma to soft tissue. It has been reported that crush syndrome accounted for more than half of the thousands of hospitalizations that occurred after the Armenian earthquake. As rescuers continued their efforts to free the victim, they realized that she was using her own psychological treatment by discussing every current issue continuously while hiding the fact that she knew her injuries were serious.

During the last phase of the rescue operation and just prior to removing her from the debris entrapping her, one of the rescuers asked if she was married. She answered that she wasn't but that she had plans to marry in the near future—which now would have to be put off for a while. The rescuers asked why she would want to put off such a happy occasion. Her answer was quick and direct: "I want to dance at my wedding." It was 18 months after the building collapse and after many hard hours of therapy that she walked down the aisle. She invited her rescuers to the wedding and danced with them.

Proper and prompt treatment of victims is essential to their survival (Figure 19.5). It's extremely important, therefore, that rescuers follow some basic rules to help ensure that victims are given the proper treatment and care, especially those involved in complex rescue operations. (*Note:* Psychological treatment can be rendered by anyone. No certifications or degrees are necessary—just good old common sense and compassion.)

The basic rules include the following:

- Ideally rescuers should be trained in the skills necessary to treat accident victims: administering CPR; treating for shock, controlling bleeding; and completely packaging the victim, including splinting, bandaging, and backboarding.
- First responders who do not have emergency medical technician, advanced medical technician, or paramedic qualifications should understand the responsibilities of these personnel and work with them, not against them. This can be facilitated by cross-training between emergency medical service and rescue personnel.
- Victims must be protected so that they do not suffer additional injuries during the rescue operation (Figure 19.6).

Figure 19.6. *Victim is removed after proper packaging.* COURTESY S. SPAK

- Victims always should be treated for possible spinal injuries: They must not be moved until qualified medical personnel have examined them and taken the proper steps to prevent further injury.
- Victims of prolonged entrapment, such as those in building collapses and industrial accidents, are prime candidates for crush syndrome. Having qualified medical personnel at these incidents is a must.

Figure 19.7. Rescuers with emergency medical service training help ensure that the victims will not suffer additional injuries. COURTESY T. McCARTHY

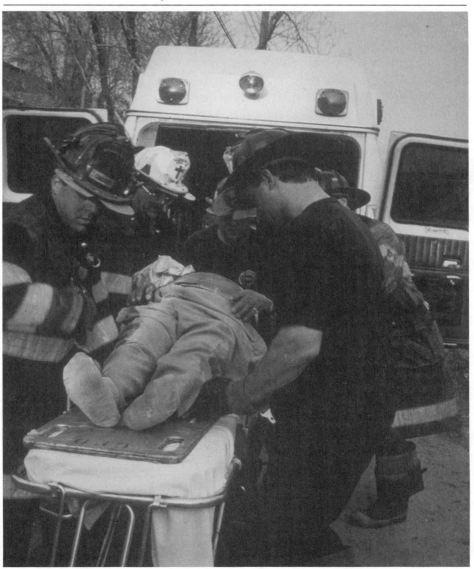

Crush Syndrome

What is crush syndrome? Why does it kill? These questions were answered by Dr. Joseph A. Barbera, the physician commander of the Special Medical Response Team that has responded to and treated the victims of the earthquakes in Armenia and the Philippines.

Dr. Barbera states: "Simply put, this condition develops when muscle tissue is compressed for enough time (usually four to six hours) so that it begins to die. When the trapped area is released, blood rushes into the impaired muscle tissue, and the plasma portion leaks into the tissue. This may occur rapidly enough and with enough blood volume that the patient develops shock and dies. Blood returning from the injured area carries multiple toxins that may cause the heart to stop (asystole) or fibrillate (v-fib), either of which results in sudden death. A later sequela in survivors is kidney failure caused at least in part by a protein called myoglobin, released into the bloodstream from injured muscle." Many of these problems can be prevented by timely medical intervention, says Dr. Barbera. This treatment should begin when the victim is initially reached by rescuers. Treating a victim after removal from the rubble may be too late.

20

CONCRETE

CONCRETE IS A building component that figures prominently in our fire-fighting and rescue operations. Our experience with concrete runs the gamut from making openings for hose streams to the much larger rescue operations involving tons of concrete. At times it is necessary to breach walls to reach trapped victims (Figure 20.1). Access openings in piers or wharfs during fires usually are required for under-pier nozzles or distributors, which provide a very effective means for reaching fire not accessible from above.

Cellar and subcellar fires can create unique firefighting problems. Just advancing into a cellar and its fire usually pushes firefighters to the extreme; and many times when they get there, they find that the fire is still a level below. Incident commanders usually order openings made in the concrete flooring for the use of cellar pipes, distributors, bent-tip nozzles, and so on. Breaching side walls is another tactic used to reach inaccessible areas. Tactics used in high-rise incidents often require that openings be made through concrete flooring for ventilation purposes and hose-stream penetration. Francis L. Brannigan, the noted expert on building construction as it involves the fire service, points out some very important facts in his book *Building Construction for the Fire Service* (National Fire Protection Association): Conventional reinforced concrete, where the tensile strength is provided by steel reinforcing rods embedded in the concrete, usually doesn't present any problems except that of having to cut the rods after breaking through the concrete. Importantly,

Figure 20.1. *A jack hammer is used to breach a building wall during a fire operation.* COURTESY W. FUCHS

Figure 20.2. *It is recommended that fire departments keep permanent records of post-tensioned buildings.* PHOTO BY AUTHOR

the potential for a collapse is not present when dealing with the conventional reinforced concrete. This is not the case with post-tensioned concrete.

Post-Tensioned Concrete

In post-tensioned concrete, the tensile strength is provided by steel cables under tremendous tension. These cables are not in intimate contact with the concrete. Brannigan notes that cutting one of these cables would be the same as cutting any steel cable under tension. The ends could whip with a tremendous energy release, and cutting the cable could cause a massive collapse. In buildings being constructed with post-tensioned concrete, the entire weight of a floor or beam rests on the formwork until the tensioning (using hydraulic jacks) is completed. Because of this, Brannigan points out that post-tensioned concrete buildings under construction present a special hazard of massive collapse. In addition, he recommends that fire departments keep permanent records of post-tensioned buildings in their districts because of the serious potential for collapse they present (Figure 20.2).

Structural-Collapse Incidents

One of the most challenging rescue operations is structural collapse. According to the American Society of Civil Engineers committee on damaged and failed structures, at least 500 large building mishaps have occurred every year for the past 10 years, many of which involved concrete structures. Fortunately, not all involved such a high death toll as the following major incidents:

- In the Oakland, California, area, firefighters responding to the collapse of Interstate 880 during the October 1989 earthquake found that almost one and one-quarter miles of the freeway's upper deck had collapsed into the lower section. Rescue workers teamed up with construction workers and spent five days clearing tons of concrete rubble (Figure 20.3).
- In the December 1988 earthquake in Armenia, which left more than 25,000 dead and demolished entire towns, rescuers faced a similar challenge: tons and tons of concrete rubble with numerous victims trapped underneath.

Figure 20.3. *Firefighters teamed up with construction workers using heavy-duty concrete breakers to penetrate concrete roadway on the Oakland, California, freeway after an earthquake in 1989.* COURTESY M. NAVARRO

- The collapse of the "L'Ambience Plaza" high-rise apartment building under construction in Bridgeport, Connecticut, in April 1987 killed 28 construction workers; rescuers spent 10 days digging through hundreds of tons of concrete and steel.
- Another high-rise construction collapse of eight-inch flooring occurred in Fairfax County, Virginia, in March 1973, killing 14 workers.
- One of the most tragic incidents involving concrete occurred at the Hyatt Regency Hotel in Kansas City in July 1981: 114 people were killed and 200 injured when a 145-foot long "sky bridge" at the fourth-floor level collapsed onto a similar sky bridge at the second-floor level, sending both crashing down onto a crowded dance floor. It took Kansas City rescuers 12 hours to dig out the victims from beneath the concrete rubble.

Tools and Equipment

How can rescuers deal with the problems presented by concrete? What tools and equipment are necessary for dealing with these incidents? In March 1989 a program sponsored by the Urban Search and

Rescue Working Group in association with the National Association for Search and Rescue was held at the University of California in San Diego. Rescue workers, municipal workers, and manufacturers tested and evaluated equipment for cutting and lifting concrete slabs. Jack hammers, demolition hammers/drills, saws and cutting equipment were tested on concrete slabs 14 inches thick, two feet wide, and nine to 12 feet long. The results varied according to the type of equipment used (large commercial equipment was compared with small portable types used by rescuers). Depth penetration, speed, and ease of handling also were considered (Figure 20.4).

Figure 20.4. *Power saw with concrete-cutting blade is used to breach wall.* COURTESY W. FUCHS

Figure 20.5. FROM LEFT: *Gasoline-operated, hydraulically operated, electrically operated, and pneumatically operated jack hammers.* PHOTOS BY AUTHOR

What is the best type of tool/equipment for a unit? That depends mainly on the type of response area and the construction features of the buildings in the area. The following factors also must be evaluated: What other equipment carried on the apparatus is capable of doing this type of work? How much space is available on the apparatus? (Large jack hammers and generators or compressors can take up a lot of space.) How big is the department's budget allocation for this type of equipment?

The tools/equipment available capable of penetrating any type of concrete come in many sizes, shapes, forms, and with various power sources. For example, *a jack hammer* (pavement breaker/drill) can be powered electrically, hydraulically, pneumatically, and by gasoline (Figure 20.5). Most utility companies and municipal departments use *large air compressors* that are transported to the job site by trailers or trucks. This provides a power source close to the operational point, cutting down on any power loss. Obviously, this type of equipment has its limitations as to where it can be used.

A unique *pneumatic breaker* on the market weighs only 11 pounds and is designed to operate from any air source (from a large compressor to an SCBA cylinder. It delivers 40 to 250 psi, depending on the material to be penetrated and operates continuously for seven minutes from a full 30-minute air-pack cylinder. It can be used to breach walls or penetrate concrete floors.

Electrically operated pavement breakers/drills powered by generators or directly from 120-volt AC can be used in cellars or upper floors of a high-rise building. Because of the versatility of electrical and small portable pneumatic breakers/drills, they are prevalent in the fire service. These tools are easier to set up, quieter, less fatiguing, and cleaner to operate. They are available in different sizes (overall weight and power capabilities). The combination breaker/drill can be used to breach or break concrete effectively first by drilling a series of holes and then switching to breaking the concrete.

Hydraulically operated hammers require power sources that at times can be cumbersome and require additional manpower to get the system into operation, especially in hard-to-reach areas.

Gasoline-powered hammers are a multipurpose tool that can provide combination drilling and breaking with very effective efficiency, but, again, they can be used only in a limited number of areas. The exhaust fumes from the tool must be considered, especially if work has to be performed in below-grade areas or indoors, where the tool operator is directly exposed. Manufacturers now have provided exhaust hoses for these machines to eliminate the problems presented by the fumes.

Figure 20.6. A hydraulically operated diamond-segmented chain saw is one of the latest state-of-the-art tools for penetrating concrete. COURTESY J. NORMAN

Figure 20.7. This amfire drill, powered by water, can be used also as an extinguishing agent. COURTESY J. SKELSON

Power saws using either silicon carbide blades or diamond tip blades are capable of cutting concrete. Also available is a *hydraulically powered chain saw* that uses a diamond chain saw blade that can cut concrete, brick, stone, and other similar materials (Figure 20.6). In the California test, a *ring saw* that uses an off-center blade-drive system was used. The system has a 14-inch blade that cuts to a depth of 10 inches.

Another unique tool available for use on concrete is a *water drill* that uses water from a pumper as the power medium for the drill bit and as an extinguishing agent (Figure 20.7). It is attached to a one and one-half inch line; when operating at a pump pressure of 200 psi, the diamond core bit cuts all masonry products including reinforced concrete.

The City of New York Fire Department uses a *concrete core cutter* designed originally for use in high-rise buildings. Electrically powered from a house current or generator, it uses an eight-inch diameter diamond core bit (12 inches long) that can cut through 48 inches, using extensions at the rate of one inch per minute. Using a telescoping floor-to-ceiling column support, it can be used to cut overhead through a ceiling, laterally through a wall at 90 degrees, or downward through the floor. Because of its size and weight, it is mounted on a wheeled platform and can be moved easily from one location to another.

Figure 20.8. *At times, good old manual labor is used to breach walls.* COURTESY W. FUCHS

Ideally, rescue companies should carry a variety of these concrete-cutting tools/equipment. Alternative power sources must be considered. At some incidents, compressor-supplied or fuel-supplied tools can be used. In other incidents, because of inaccessibility, confined space, or other reasons, a small portable pneumatic tool may be the only one that will do the job (Figure 20.8).

As in all rescue operations, it is necessary to drill and train with tools and equipment so that the rescuers become familiar with them, their capabilities, and their limitations. It must always be remembered that the scene of an operation is not the time to test equipment.

21

PLANNING FOR MAJOR OPERATIONS

COURTESY S. SPAK

"See you at the big one" or "See you on the 11 o'clock news." are two of the expressions firefighters use instead of the traditional "goodbye." A fifth-alarm spectacular warehouse fire or the dreaded tragic loss-of-life fire usually constitutes some of the "big ones" or major operations for firefighters. The "big ones" for firefighters assigned to the rescue company most likely would include building collapses, air crashes, train accidents, and derailments, fires, hurricanes, tornadoes, floods, high-rise fires, large-scale hazardous-materials incidents, and the type of disaster that has had widespread media coverage, earthquakes. Many departments have developed standard operational procedures (SOPs) for various fire operations that may include private and multifamily dwellings, taxpayers (attached commercial stores), shopping malls, or the most prevalent type occupancy or building construction (frame or heavy mill, for example) in the response area.

Plans for major operations should be formulated in accordance with SOPs, with one exception: They should be generic and capable of being adapted to the type of incident rather than be designed for specific objectives (Figure 21.1). Building-collapse and air-disaster plans, for example, would use the services of a heavy rescue unit; but the initial priority duty assignment at each incident may differ with regard to immediate rescue. The plan design would be the same format, i.e., incident command system, chain of command, and so forth; but the actual execution can vary, understandably, according to the type of incident.

Figure 21.1. *Heavy rescue assignments may differ in nature, but their major operational plan is the same.* COURTESY CITY OF NEW YORK FIRE DEPARTMENT

Initial Steps

The major operational plan can be implemented in a timely and appropriate manner by initiating the following actions at the onset of the operation:

- Implementing a response assignment capable of handling the incident at least during the early stages of the operation
- Establishing a notification signal that dictates a specified response that includes the firefighting and special units needed for a specific major operation
- *Establishing a code to be used in conjunction with the notification signal that indicates the apparent seriousness of the incident or alerts additional units to assume standby status*
- Using the signal *10-99* followed by the code 1 (10-99, code 1), for example, would indicate that a major operation response is required at the incident. This type of proaction ensures a response that will meet the incident commander's needs. If the assigned units are not needed, they easily can return to their stations.

A signal *10-99, code 2* would indicate a major operation incident less serious in nature that necessitates an initial, but reduced, response assignment; it also would relate that additional units may be required and must be in a standby position should the incident commander upgrade the code 2 to a code 1.

- *Code 3* could be used for incidents less serious in nature than code 2 that have the potential to escalate and require a code-2 or a code-1 assignment. The code-3 signal serves as a notification to all affected units and places them in a readied state.

The signal *10-99* should be followed by a brief description of the type of incident. As an example, after transmitting the signal *10-99, code 1*, the following would be added: "We have a serious building collapse with possible trapped occupants." Or, the signal *10-99, code 2* could be transmitted and then the message, "We have a derailed work train and are investigating to see if there are any injuries." These details should be included when possible; but the message should be brief so that the dispatcher can notify all affected units. Urgent messages should be transmitted immediately.

Proacting in this way—instead of merely reacting to the developments of the situation—eliminates the need for the incident commander to have to wait for the additional requested troops.

- *Implementing the incident command system for all major operations.* Most departments refer to the incident command system (ICS) as a management tool in that it provides the incident commander with an effective means for managing and controlling an incident regardless of its size. The system identifies definite lines of authority and a command structure that gives the incident commander the capability to manage the overall incident while those delegated under the system, and who are directly responsible to the incident commander, manage and control the many segmented areas of responsibility. Discipline is a key element of every major operational plan, and an incident's magnitude shouldn't interfere with the control and discipline that must be maintained to control a major operation (Figure 21.2). "Freelancing" (performing tasks without orders and without knowing the overall game plan, which results in the firefighters' failure to fulfill the instructions or duties assigned to individual members or the unit) must be avoided; adhering to ICS guidelines helps ensure that the major operational plan will accomplish its goal.

In a major operational plan, the notification signal and its code indicate the type of incident and the degree of seriousness, its scope, and its complexity. A major operational plan is based on

Figure 21.2. *Discipline is a key element of every major operational plan. Members delegated and charged with responsibilies are answerable to the incident commander.*
COURTESY W. FUCHS

department resources, interagency agreements, mutual-aid considerations, and often on prior experience.

The Major Operational Plan and a Building Collapse

Building collapse have been occurring more often than in previous decades and by all indications will increase, presenting many challenges for rescue units. The problems presented by building collapses generally are related to factors such as the type of construction, the area of collapse, the height of the collapsed structure, and the occupancy of the building. Would the same response be appropriate for the following two building-collapse scenarios: the 4 a.m. collapse of a one-story vacant private dwelling with no exposure problems in a desolate area of the city and the collapse of an occupied six-story brick-and-joist multioccupancy commercial building on a heavily traversed street during a workday lunch hour?

Scenario A

The officer transmitting the notification signal for the one-story building most likely would opt for the less serious code, while the officer first arriving at the lunch-hour disaster undoubtedly would transmit the code for the most serious type of indecent (Figure 21.3) (and probably in a louder-than-normal tone, which clearly signifies a serious incident).

Each incident certainly presents different needs, which are built into the appropriate codes. The code for the less serious incident includes the normal first alarm firefighting complement assigned to the street location, a senior chief, a safety officer, a rescue company, a special collapse or cave-in unit, lighting apparatus, and any other units requested by the incident commander.

Figure 21.3. LEFT: *A less serious incident generally necessitates a different response and notification signal than (right) a larger disaster.* PHOTOS COURTESY AUTHOR AND J. IORIZZO, RESPECTIVELY

Figure 21.4. *A number of additional units—more heavy rescues, special collapse vehicles, support units, and other appropriate equipment—would respond to a more serious incident.* COURTESY H. EISNER

Scenario B

The more serious incident obviously demands more resources, and the coded signal would provide for the necessary additional units that could include the following:

- The normal first-alarm assignment, including a snorkel (also known as a cherry picker, an elevated platform, or a tower ladder/platform)
- Additional rescue companies—three or four rescue companies may be needed in some cases (Figure 21.4).
- Special collapse, cave-in, or other designated units that carry the specialized tools/equipment needed for these incidents
- Specialized support units that may provide lighting, generators, compressors, and other equipment

- A hazardous-materials unit
- Additional chief officers to include a superior chief officer and enough chief officers to staff an expanding ICS
- A safety officer
- A communications unit capable of handling the increased radio, handie-talkie, and telephone traffic
- A photo or video unit
- The presence of other agencies such as the police department (for traffic and crowd control), emergency medical personnel, utility companies (gas, electric, water, and telephone), the building department, and any other agency that could provide a service needed during the operation.

The plan must provide for the tactical placement of the first-arriving units. A snorkel must be positioned as close as safety permits to provide a means for observing and surveying the collapsed area (Figure 21.5), or to provide master-stream protection for rescuers, if needed. The first and second engine companies must be positioned to provide handline

Figure 21.5. A snorkel can be used as an observation post from which the collapsed areas can be surveyed. COURTESY H. EISNER

protection (Figure 21.6). The first arriving rescue company should be positioned as close to the incident as possible; all other units should be directed to an apparatus staging area. Positioning apparatus in this manner keeps them out of the collapse zone/area, reduces apparatus noise and vibrations during victim search and rescue, keeps the area clear, and provides access for specialized equipment that may be needed close to the collapse area (heavy equipment, cranes, additional snorkels, additional lighting equipment). Vehicles such as civilian cars or trucks blocking access lanes or interfering with the positioning of apparatus should be towed. The first and second engine companies provide handline protection and backup lines for rescuers, if needed. A third engine company should proceed to the rear of the building or the most serious exposure problem or the location directed by the incident commander. Handline protection must be provided regardless of the circumstances; stretching lines after an explosion oftentimes is too little too late. The first ladder companies will conduct an exterior survey and search the collapsed area. One of the first arriving ladder companies must make sure utilities are shut down. Collapses can cause broken gas or water pipes and severed electric lines, which, if not identified and corrected during initial sizeup operations, could cause disastrous secondary events.

The first arriving chief officer or incident commander will have his/her hands full and the "computer" (under the helmet) will be working, possibly in overload. The reports transmitted by units on the scene will give the IC an idea of the extent of the incident. The IC, upon arrival, immediately should announce his/her presence and identify the location of the command post (Figure 21.7). At a recent major fire, the IC failed to announce the location of the command post and had positioned himself at what he believed to be the front of the fire building. The fire reached a point where it was burning on two street fronts. Most first arriving units were operating on one street, and the chief was on the other street. Trying to explain how the fire travel was advancing became confused since the chief had a different perspective of the incident.

After setting up the command post, the IC must determine from the units on the scene the actions taken prior to his/her arrival and the present status and conditions of the incident. A size-up, additional orders, and the locations of the command post and the staging area must be given to the dispatcher for relay to units en route. Pending arrival of the additional chief officers, the IC must prepare to implement the ICS and issue the necessary orders and instructions to arriving officers. Rescue companies at the scene generally provide the IC with information and updates regarding survey and reconnaissance of the collapsed area, possible voids, and accessible areas for victim search and rescue. Only

Figure 21.6. The engine company providing handline protection is out of the collapse zone. PHOTO BY AUTHOR

Figure 21.7. The incident commander must identify the location of the command post upon arrival at the scene. COURTESY S. SPAK

Figure 21.8. *Rescue personnel must be skilled at tunneling, trenching, and/or cutting through collapse debris.* COURTESY S. SPAK

those trained in this type of operation should enter these voids and accessible areas. Tunneling, trenching, or cutting through collapse debris requires the expertise of rescue personnel equipped, trained, and experienced in collapse search, rescue, and removal procedures (Figure 21.8).

These specially trained rescue personnel use a system designed to cope with the many difficult and unusual situations resulting from building collapses. The "six-sided" approach rescuers use when implementing their rescue plan involves strategically positioning the units to approach each incident situation from above, below, and each of the four sides, (usually referred to as exposures 1, 2, 3, 4, or sectors 1, 2, 3, 4) (Figure 21.9). Looking from the command post located in front of the incident site, the exposure would be as follows: 1–the area directly in front of the building/site, 2–directly to the left, 3–to the rear, and 4–the right side. Using the standard terminology for the these operations ensures that all personnel understand the issued instructions, directions, and orders. A major operation plan must provide guidelines that ensure a safe operation for rescuers and victims.

Figure 21.9. *Exposures.* COURTESY AUTHOR

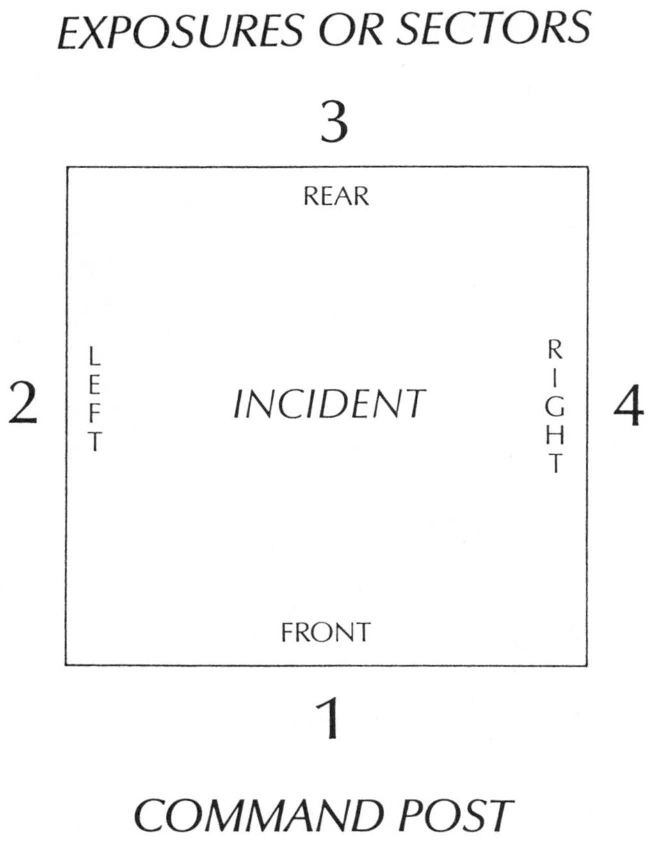

EXPOSURES OR SECTORS

The Plan By Stages

Working the major operational plan by stages facilitates managing and controlling the incidents. The following stages apply to building collapses:

STAGE 1

Rescuers survey and do a reconnaissance of the entire area (Figure 21.10), looking for trapped victims and assessing the potential danger for rescuers. Shutting down all utilities (gas, electric, water) where possible eliminates some of the sources that could cause injuries to rescuers or victims. Rescuers should not compromise their safety by attempting to shut down utilities. If the utilities cannot be shut down without exposing rescuers to dangers, the incident commander must be notified. The utility companies generally will assist in the shutdown. All personnel must be notified if utilities are not shut down during the rescue operation.

Figure 21.10. During Stage 1, rescuers survey and do reconnaissance of the entire area.
COURTESY T. McCARTHY

STAGE 2

Surface victims are removed as quickly and as safely as possible. The rescuer must be extremely careful during this stage. Rubble piles resulting from collapses often are very unstable, and their outer appearance can fool rescuers. What appears to be a settled pile of debris on the surface in reality could be totally unsupported from below. Secondary collapses also can occur during this stage.

Caution is needed also when approaching the victim. Removing obstructions while advancing toward victims must be done cautiously, to prevent an additional or secondary collapse.

STAGE 3

All voids and accessible spaces created by the collapse structure must be searched for possible victims (Figure 21.11). Rescuers entering these voids can use the call-and-listen method (as previously explained: the rescuer calls out and waits for an answer) in their attempts to locate victims. If an answer is received, rescue operations are directed toward the location from which the sounds are coming. Depending on the size and area of the spaces, the search can be a very time-consuming operation. To emphasize again: Only specially trained rescue personnel should operate in these voids and spaces. Their training and experience ensure a safe and successful removal procedure. Shoring and bracing may be required during this stage to prevent additional collapsing or the shifting of debris and possibly serious injury or death to victims and rescuers. It can't be stressed enough that the personnel in charge of this type of operation be experienced and have a background that includes collapse-rescue operations. Rescuers' safety is always a prime consideration. When necessary, implementing this stage may have to be delayed until dangerous or unsafe obstructions are removed.

STAGE 4

Selected debris is removed (Figure 21.12). At times, the size, area, and condition of collapsed structures themselves can be the reasons for proceeding to this stage. After locating victims, it may be necessary to remove obstructions that are impeding the operation. Selected debris removal usually is instituted after receiving information pertaining to victims' locations before the collapse. This information results in a more concentrated rescue operation at a specific location instead of in a num-

Figure 21.11. *All voids and accessible spaces must be searched and explored for possible victims; this is done during Stage 3.* COURTESY J. IORIZZO

Figure 21.12. *Selected debris is removed during Stage 4.* COURTESY S. SPAK

ber of simultaneous operations covering a wider area. The concentrated search conserves manpower. Secondary collapses, again, must be considered during this stage.

STAGE 5

General debris is removed (Figure 21.13). This usually is done after all victims have been removed and accounted for. The main consideration during this stage is determining whether additional victims—the "what ifs"—visitors to the collapsed structure, such as delivery and post office personnel—are in the debris. During this stage, heavy equipment is used to clear the rubble. If any possibility exists that additional victims may be in the pile, debris should be removed with extreme caution. The IC decides when heavy equipment is to be used for removing debris. All rubble removed must be searched, recorded, and removed to a predetermined location and remain there until the responsible agency authorizes its removal. The nature of the incident may have legal ram-

Figure 21.13. General debris removal, done during Stage 5, usually requires heavy equipment. COURTESY S. SPAK

Figure 21.14. *Incident command chart.*

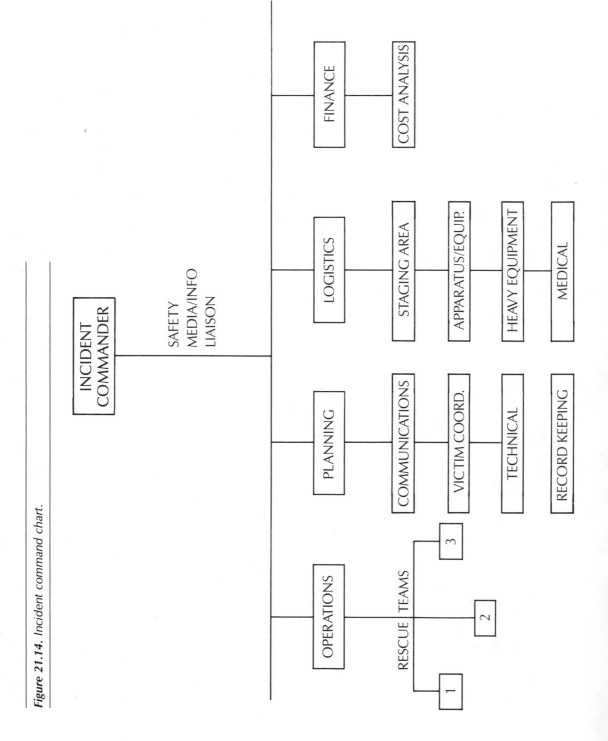

ifications that must be considered. To manage and control the situation effectively, an incident commander must use all components of the incident command system during major operations (Figure 21.14).

The operations commander is responsible for the tactical operations of the rescue teams. Depending on the size of the operation, the operations officer may have to enlarge the number of teams for effective search, rescue, or victim removal. Although the officer has to delegate some of his duties to cover the enlarged team operation, the responsibilities remain the same, and the IC must be kept informed and updated as often as necessary.

Major Operational Planning

The *planning section commander* provides the incident commander with the status of resources and components of the operation. Communications and information, which flow vertically and laterally, includes reports from technical specialists regarding the building's construction, environmental concerns, and specialized operations such as hazardous materials, victim-tracking, and coordinating and documenting the entire incident. The responsibilities assigned to the planning section commander are not completed until all units are demobilized and all resources released from the incident.

The functions assigned to the *logistics commander* include providing support services and facilities during the incident. These services include ensuring that the medical components provide the necessary amount and level of medical aid and transportation for all victims and personnel, assist in the operations of the command post and staging areas, and provide rest and rehabilitation zones for personnel. The logistics commander is responsible for providing canteen services or other means to feed personnel during an extended operation. The logistics commander also might be assigned additional functions such as servicing and repairing apparatus, providing special equipment needs, and replacing essential tools and equipment.

A *finance commander* is responsible for collecting data during the incident that pertains to cost estimates, cost recovery, and budget justification. The possibility of cost recovery always must be considered, especially when the incident involves a governmental or other agency that is responsible for reimbursing the operating department by virtue of a pre-incident agreement or arrangement for the services. Analyzing operational expenditures can provide the data needed for future major operational planning.

Figure 21.15. *The safety officer is directly responsible to the incident commander.* COURTESY J. IORIZZO

The *safety officer* (Figure 21.15), *information officer*, and *liaison officer* are directly responsible to the incident commander and provide the means by which the important areas of responsibility included in the ICS are managed and coordinated directly with the incident commander without interrupting the commander's overall management and control of the incident. The handling of important functions by other officers—such as interagency coordinator, which is handled by the liaison officer—allows the incident commander to concentrate on his primary responsibilities without distractions and yet be fully cognizant of the status of the incident. The ICS is a vital part of major operational planning. Delegating authority through the use of the ICS provides greater control and accountability of personnel, tactical effectiveness, and definitive lines of responsibility. Although the vertical structure does not show direct horizontal communication, it must be remembered that all communications should be lateral in addition to vertical for purposes of coordination and safety. Decisions affecting the strategy and tactics of an operation must be clearly understood by all personnel involved. A major operational plan can involve many separate and distinct parts; using the ICS with a well-coordinated and well-executed communications flow ensures success.

Experience has shown that no two incidents are exactly the same. Trying to prepare and train for every possible scenario would take up most of our living hours, and then some. What we can do, however, is develop major operational plans that can be adapted for a variety of possible incidents, plan drills using the major operational plan, and improvise and adjust as the scenario plays out. Why improvise and adjust? Earlier in this chapter was discussed a prearranged notification signal that uses a code to relay the varying degrees of seriousness associated with incidents (Figure 21.16). The examples of the two collapsed buildings that required different responses can apply to other types of major incidents as well: a large aircraft disaster involving a commercial aircraft in an urban setting versus the crash of a private two-seater craft in a cornfield, a rail disaster that involved a fully occupied "overnighter" train colliding with a highly toxic chemical railcar versus two work trains in a rail yard colliding, for example.

The major operational plan can be used regardless of the type or seriousness of an incident. Developing major operational plans and adapting and adjusting them up and down to fit the circumstances ensure effec-

Figure 21.16. *The size of the incident influences the type of notification signal used in the major operational plan.* COURTESY S. SPAK

Figure 21.17. *To provide an efficient and organized system for managing an incident successfully, the major operational plan must be rigid and yet flexible enough to be adapted to the complexity and uniqueness of the incident.* COURTESY H. EISNER

tive and successful operations (Figure 21.17). Managing and controlling incidents will be accomplished by using the incident command system as a vital component of the major operational plan. An effective plan, as already explained, should be rigid in its lines of authority and structure and yet be flexible enough to be adapted to the complexity and uniqueness of the incident without compromising the guidelines and concepts necessary for an efficient, organized, and successful system.

INDEX